SpringerBriefs in Earth System Sciences

SpringerBriefs Seaways and Landbridges: Southern Hemisphere Biogeographic Connections Through Time

W0079787

Series Editors

Kevin Hamilton
Gerrit Lohmann
Lawrence A. Mysak
Justus Notholt
Jorge Rabassa
Vikram Unnithan

For further volumes:
http://www.springer.com/series/10032

Susana E. Damborenea
Javier Echevarría · Sonia Ros-Franch

Southern Hemisphere Palaeobiogeography of Triassic-Jurassic Marine Bivalves

 Springer

Susana E. Damborenea
Departamento Paleontología Invertebrados
Museo de Ciencias Naturales La Plata
La Plata
Argentina

Sonia Ros-Franch
Departamento Paleontología Invertebrados
Museo de Ciencias Naturales La Plata
La Plata
Argentina

Javier Echevarría
Departamento Paleontología Invertebrados
Museo de Ciencias Naturales La Plata
La Plata
Argentina

ISSN 2191-589X   ISSN 2191-5903 (electronic)
ISBN 978-94-007-5097-5   ISBN 978-94-007-5098-2 (eBook)
DOI 10.1007/978-94-007-5098-2
Springer Dordrecht Heidelberg New York London

Library of Congress Control Number: 2012944966

Printed on acid-free paper

Springer is part of Springer Science+Business Media (www.springer.com)

Preface

The study of global biodiversity changes is a strong issue these days, as we become aware of the fragility of the Earth system and the urgent need to understand it better to keep it working. One of the key aspects of biodiversity is the distribution of organisms, and biogeography is the discipline which tries to recognize and characterize the causes of that distribution. It is closely linked to ecology, since the distribution of organisms is related to ecologic factors, but it cannot ignore other matters, such as species origin, dispersal, and extinction, and thus it becomes a historic science. Introducing the important time dimension, scientists are turning their interest to the past distribution of organisms, and paleobiogeography is now a complex subject which processes information provided by both Biology and Earth Sciences. It is conceptually and philosophically equivalent to neobiogeography. Nevertheless, its methods are somewhat different, since it is seriously limited by the incompleteness of the fossil record. On the other hand, it has direct access to the time involved, a key ingredient of organic evolution.

This book is a synthesis of many years of research on Mesozoic bivalve mollusks from South America. The task of updating and processing all the information was triggered by a Symposium on "Seaways and landbridges: Southern Hemisphere biogeographic connections through time", organized by Dr. Silvio Casadío (CONICET and Universidad Nacional de Río Negro) and Dr. Miguel Griffin (CONICET and Universidad Nacional de La Plata), held at the Universidad Nacional de Río Negro, General Roca, Río Negro Province, Argentina, April 26–May 1, 2011, and who kindly encouraged us to complete this contribution. After that memorable meeting we put together our knowledge and skills to produce this book together, and present a picture of present-day knowledge on Triassic and Jurassic bivalve distribution in the Southern Hemisphere. Benthonic bivalves have been widely used in biogeography and paleobiogeography, but data from the Southern Hemisphere Mesozoic are still few and scattered. We trust this book will encourage future research on this subject to build better databases and allow the application of more sophisticated methodology. Through the book we analyze the subject from several viewpoints. From a merely descriptive perspective, bivalve distribution shows latitudinal gradients and distributional patterns, both at regional and global scales, which are the basis for the recognition of biochoremas (i.e., paleobiogeographic units of different ranks). Moving forward toward a causal paleobiogeography, these organisms also provide interesting insight into particular biogeographic questions, such as bipolarity and its origin. The evolution in time of the recognized biochoremas can be discussed in relation to paleoclimates and paleogeography. Finally, some of the results obtained from the analysis of the distribution of past bivalve biotas were even used to propose and discuss the development of marine corridors and argue about the distribution of continents in the past.

Susana E. Damborenea
Javier Echevarría
Sonia Ros-Franch

Contents

Chapter 1
Introduction

Abstract Bivalves have proven to have a great potential for paleobiogeographic analyses due to their relatively complete fossil record, especially for Mesozoic and Cenozoic times. Being mostly benthonic, they have a large variety of life habits which should be taken into account, particularly in detailed paleobiogeographic studies. We will analyze marine bivalve distribution in the Southern Hemisphere during several successive time slices within the Triassic and Jurassic, an epoch marked by critical geologic and biotic events. This period covers both the biotic recovery after the harshest diversity crisis ever (the Permian/Triassic extinction event), and later also the biotic reaction to another severe crisis at the Triassic/Jurassic boundary. This allows the opportunity to evaluate the response of paleobiogeographic patterns to such events. The Earth's configuration drastically changed from a concentration of land masses in a unique supercontinent (Pangea) and two oceans (Tethys and Panthalassa), to a fragmented series of continental land masses. These began to disperse, opening sea corridors which largely affected not only the global distribution of biotas but also paleoclimate and sea paleocurrents as well. This dynamic paleogeography adds an interesting ingredient to the study of past distributions of benthic organisms making it possible to frame them into a physically and biologically changing scenario.

> *'Every naturalist who has directed his attention to the subject of the geographical distribution of animals and plants, must have been interested in the singular facts which it presents… Of late years, … a great light has been thrown upon the subject by geological investigations, which have shown that the present state of the earth, and the organisms now inhabiting it, are but the last stage of a long and uninterrupted series of changes which it has undergone, and consequently, that to endeavour to explain and account for its present condition without any reference to those changes (as has frequently been done) must lead to very imperfect and erroneous conclusions.'*
> Alfred Russel Wallace 1855

S. E. Damborenea et al., *Southern Hemisphere Palaeobiogeography of Triassic-Jurassic Marine Bivalves*, SpringerBriefs Seaways and Landbridges: Southern Hemisphere Biogeographic Connections Through Time, DOI: 10.1007/978-94-007-5098-2_1, © The Author(s) 2013

Alfred Russel Wallace, considered the nineteenth century's leading expert on the geographic distribution of animal species and sometimes called the "father of biogeography", already recognized the importance of studying the history of biotas long before moving continents and plate tectonics were heard or even thought of. On the other hand Darwin, who initially recognized the importance of geographic isolation to speciation in his unpublished notebooks (see Lieberman 2003, 2008), did not mention this in his later publications.

The study of global biodiversity changes is a hot issue these days, as we humans become aware of the fragility of the Earth system and the urgent need to understand it better to keep it going. One of the key aspects of biodiversity is the distribution of organisms, and *biogeography* is the discipline which tries to recognize and characterize the causes and patterns of distribution. Biogeography is closely linked to ecology, since the distribution of organisms is governed by ecologic factors, but it cannot ignore other matters, such as the origin of species, their dispersal, and extinction, and thus it can be considered a historic science.

Biologists are beginning to investigate the causes of the great global biodiversity changes that are now taking place on the Earth. But paleontologists, who possess a much more extensive time perspective, are constantly observing and surveying the changes in biodiversity produced at various times in the past, and they have the most precise access possible to this very important dimension: time. Thus, *paleobiogeography*, which studies the distribution of organisms in the past, is a very complex subject that combines information from both biology and the Earth sciences (Cecca 2002), and "paleobiogeographers can actually monitor how the Earth and its biota have evolved and coevolved" (Lieberman 2000). The data provided by the fossil record are increasingly being used in combination with other sorts of data in modern biogeographic analysis. The relationship between geology and biogeography is then unavoidable, and should be based on a reciprocal illumination approach (Morrone 2009).

Similarly to biogeography, different approaches can be recognized for paleobiogeography (Rosen 1992, 1995):

1. Descriptive paleobiogeography: recognition and description of the distribution of organisms. The outcome is the definition of biogeographic units or biochoremas. Both quantitative (more frequent in neobiogeography) and qualitative (more subjective) methods can be used.
2. Causal paleobiogeography: examines the causes of the observed distributions. There are many arguments related to theoretic biogeography and the philosophic approaches, which will not be discussed further here (for a good synthesis see Cecca 2009). According to the temporal scale of the processes involved, two main viewpoints allow distinction between ecologic biogeography, with a temporal scale related to biologic cycles, and historic biogeography when long-term processes are analyzed.
3. Applied paleobiogeography: the distribution of organisms can be used to infer the role of ecologic factors, the relation between phylogeny and provinciality, or paleogeographic patterns.

Through the book we will follow all these three approaches using Mesozoic bivalves from the Southern Hemisphere, and try to apply the resulting knowledge to the discussion of regional and global issues. Since scale is a key element in paleobiogeography, the chapters will be ordered progressively according to the regional (Chap. 4), hemispheric (Chap. 5) and finally (Chap. 6) some global issues involved.

During a long time of research on the distributions of Mesozoic bivalves, we became increasingly aware of the huge gaps in our knowledge. At this point we fully agree with Rex et al. (2005, p. 2288) statement that "making sense of large-scale marine biogeography remains a difficult challenge", and we dare say it is certainly much harder when we plunge into deep time as we will try to do here. In this mood, we pick up the gauntlet and offer this review as a small but (we hope) significant contribution to the fascinating subject of the evolution of the Earth's biotas.

In this first chapter, the necessary framing concepts about the organisms and the time involved, some aspects on Mesozoic paleoclimate, and marine paleogeography and paleocurrents, will be laid out briefly.

1.1 Paleobiogeography and Neobiogeography

Although when we refer to the distribution of living organisms we are doing biogeography, some authors have introduced the term neobiogeography to clearly distinguish it from paleobiogeography (see Rosen 1995). These two disciplines are conceptually and philosophically equivalent, and the only important differences concern the object of study and the methodology applied, both aspects highly conditioned by the availability and limitations imposed by the data, as extensively discussed by many authors (see for instance Rosen 1992; Lieberman 2000; Cecca 2002) and summarized in Table 1.1. New techniques are continually being proposed and tested to overcome these limitations, and paleobiogeography and neo-biogeography are now converging fast into a unified discipline (Cecca et al. 2011).

1.2 Why Bivalves?

Benthonic mollusks, and among them especially the bivalves, have a great potential for paleobiogeographic analyses. They are usually considered in marine neobiogeography studies (see Flessa and Jablonski 1995 and references therein), and they were also frequently used in paleobiogeography, particularly for Mesozoic and Cenozoic times (e.g. Hayami 1961, 1984, 1987; Hallam 1967, 1969, 1977; Kauffman 1973; Zinsmeister 1979, 1982; Crame 1986, 1987, 1993; Liu 1995; Niu et al. 2011; among many others). They are generally well preserved, which is reflected in a relatively uninterrupted paleontological record, they are

Table 1.1 Some practical differences between neobiogeography and paleobiogeography

Neobiogeography	Paleobiogeography
Easier access to terrestrial environments	Mostly restricted to marine environments
Deals with barriers, isolated regions	Almost no physical barriers
The distributions can be quantified (geographical areas, abundance)	Neither the areas nor the abundance can be easily quantified
Quantitative methods are used, can be checked	Very incomplete data, hard to be checked
Molecular biology data can be used	No molecular biology data available
Geographic and environmental factors are known and can be measured	Geographic and environmental factors can only be inferred by indirect evidences
Only short processes can be studied	There is direct access to the time dimension and thus long processes can be studied
Evolutionary histories are inferred without taking into account the time dimension	The evolutionary histories are confirmed by the fossil record and are referred to the stratigraphic time scale
It is strong on processes	It is strong on patterns
It may be used to infer past geographic scenarios	It is usually used when discussing paleogeography
Difficulties arise when long-term processes are involved	Long-term processes, such as changes in distribution patterns, can be followed through time

abundant in all marine environments, they are highly diverse and have a great dispersal potential at the larval stage. Despite their age, Mesozoic bivalve faunas display a clearly modern composition, being very different to Paleozoic ones, and bivalves are one of the main components of the Modern Evolutionary Fauna in Sepkoski's (1981) sense.

It is well known that many bivalves are highly facies-dependant, and this fact should be taken into account when analyzing their distribution for paleobiogeographic purposes (Hallam 1969, 1971). Substrate type is a key factor for bivalve distribution, and it may change locally and regionally according to the bottom conditions, sediment supply, and water movement. Nevertheless, if adequately dealt with, facies dependency can turn from an apparent drawback to an asset in paleobiogeographic studies, since it allows us to interpret faunal distribution in a large range of marine habitats.

Bivalves have a wide variety of life habits, mostly related to substrate type, and these can fortunately be inferred from shell shape (see Stanley 1970). The main life habits common among Mesozoic bivalves are depicted in Fig. 1.1, and these are the categories being used throughout the text. They can be grouped or further qualified according to different criteria: (1) if using relation to the substrate into epifauna, semi-infauna, and infauna; (2) if using mobility into relatively mobile and immobile; (3) if using feeding type into detritus feeders and suspension feeders.

Also, the geographic distribution patterns of bivalves are widely ranged, from extremely local for some species to nearly global for others. Nevertheless, all authors agree that cosmopolitan species evidently dominated Mesozoic bivalve

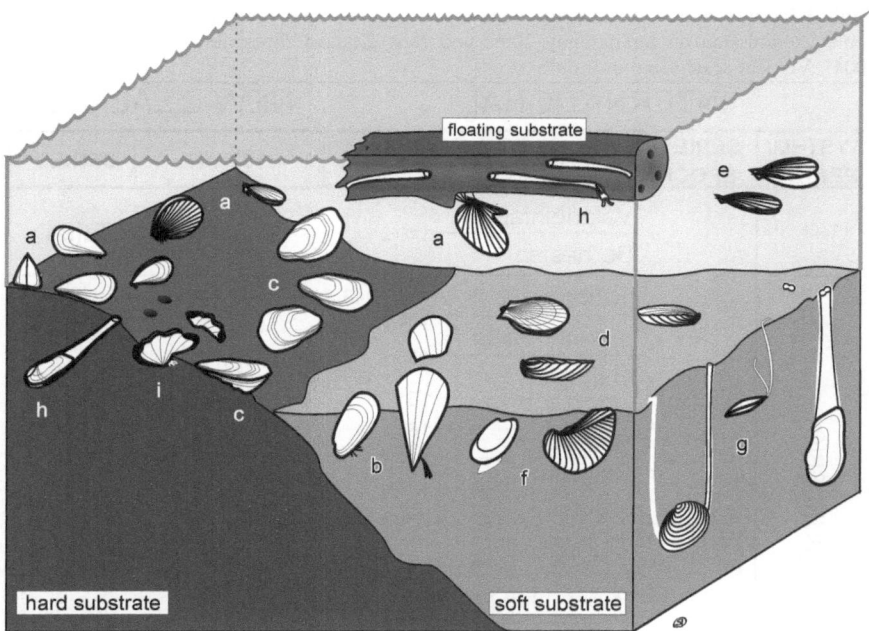

Fig. 1.1 Block diagram of a littoral marine environment showing main bivalve ecomorphologic groups represented in the Mesozoic, sketches not to scale. **a** Epibyssate (e.g. *Lycettia*, *Camptonectes*, pterioids), **b** endobyssate (e.g. *Modiolus*, *Pinna*), **c** cemented (e.g. oysters), **d** recliner (e.g. *Weyla*), **e** swimmer (e.g. *Parvamussium*), **f** shallow burrower (e.g. nuculoids, trigonioids), **g** deep burrower (e.g. lucinoids, *Thracia*, pholadomyids), **h** borer (e.g. pholadoid), **i** nestler (e.g. *Ctenostreon*). Modified from Damborenea in Camacho et al. 2008

biotas (to a higher degree than in modern biotas), at least during Triassic and Jurassic times, leaving only a few taxa to be used to define and characterize paleobiogeographic units.

1.3 Time Frame

The time frame for this book is summarized in Table 1.2. We will refer to the Triassic and Jurassic periods, which extend just after the major biotic crisis ever, the Permo-Triassic extinction event, to the end of the Jurassic, approximately between 251 and 145 Ma. This is a very interesting period of time because it shows first the dramatic recovery of biotas after the crisis, and also marks the development and establishment of modern marine faunas (Ros et al. 2012). The period is marked by critical geologic (see Sect. 1.4) and biotic events. The biota underwent several diversity crises during this interval: a severe one at the end of the Triassic and several minor ones.

Table 1.2 International stratigraphic units for the Triassic and Jurassic, according to International Commission on Stratigraphy 2009, and New Zealand timescale (according to Cooper 2004). Vertical scale represents time

	INTERNATIONAL			NEW ZEALAND	
SYSTEM/ PERIOD	SERIES/ EPOCH	STAGE	AGE Ma	STAGE	SERIES
CRETACEOUS ↑		Valanginian	140.2		Taitai
		Berriasian			
			145.5		
JURASSIC	LATE	Tithonian	148.5 150.8	Puaroan	Oteke
				Ohauan	
		Kimmeridgian	153.5 155.7		
		Oxfordian	157.5 161.2	Heterian	Kawhia
	MIDDLE	Callovian	164.7		
		Bathonian	167.7	Temaikan	
		Bajocian	171.6		
		Aalenian	175.6		
	EARLY	Toarcian	183.0	Ururoan	
		Pliensbachian	188.0 189.6		Herangi
		Sinemurian	196.5	Aratauran	
		Hettangian	199.6		
TRIASSIC	LATE	Rhaetian	203.6	Otapirian	
		Norian	216.5	Warepan	Balfour
				Otamitan	
		Carnian	227.5 228.0	Oretian	
	MIDDLE	Ladinian	237.0	Kaihikuan	
		Anisian	245.0	Etalian	Gore
				Malakovian	
	EARLY	Olenekian	249.7	Nelsonian	
		Induan	250.4 251.0		
PERMIAN ↓		Changhsingian		Makarewan	

A stratigraphic frame is preferred over the absolute dating for several reasons. Knowledge of precise absolute ages is still very unstable for this period of time and this is especially critical for the Southern Hemisphere. Consequently, instead of absolute dates, we prefer the widespread use of stages and biostratigraphically based time units, which allow the use of data compiled at different times, and are not seriously affected by new developments in absolute dating techniques. Table 1.2 shows the approximate equivalence of globally recognized stage units with those established for New Zealand, since these last have a wide use in the area (they are also applied to New Caledonia and Antarctic sequences, for instance) and thus will also be frequently referred to in this book.

Most of the analyses pursued and discussed in this book were developed within time slices coinciding with the different stages (or group of stages) of the International Stratigraphic Chart, as a way of obtaining comparable results in a time succession, thus amenable to further analysis to understand general processes. The results obtained for each of these units were then used as reference for the historic evolution of general paleobiogeographic patterns at a very broad scale. In this first approach, our time slices are admittedly very "thick" (i.e., long), and, what is worse, of very uneven "thickness" (length) if stages are used, as is evident from Table 1.2, in which the vertical scale is proportional to time in Ma. The great difference in stage duration (cf. for instance Norian and Rhaetian) introduces another distorting factor and hinders comparison. For the moment there is no way of building a comprehensive database applying time slices below the stage level.

1.4 Paleogeography

Several major geologic events with significant paleogeographic consequences took place during the time interval considered here, the main ones being the opening of the North Atlantic Ocean and the Mozambique Corridor (Fig. 1.2).

The Earth's configuration thus drastically changed from a concentration of land masses in a unique supercontinent (Pangea) and two oceans (Tethys and Panthalassa or Paleo-Pacific) at the Triassic and earliest Jurassic, first into Gondwana and Laurasia and later into a fragmented series of continental land masses which began then to disperse (Fig. 1.2b). The geodynamic processes involved are numerous and complex, and are still being the subject of research. Some of the still hotly controversial points refer to the dates of the rifting and initial splitting along plate boundaries and so-called "marine corridors" and will be further discussed later (Sect. 6.2). A huge amount of geologic and geophysic evidence was accumulated and discussed in the last few decades. In the context of this book, which deals with the distribution of (mostly shallow) marine organisms, it should be reminded that frequently the paleobiogeographic data yield older dates than the earliest demonstrable occurrence of oceanic crust, since shallow-water or intermittent connections suitable for the dispersal of littoral benthonic faunas can occur even during the initial rifting phase.

The evolution of the Earth's configuration during the time here considered was accompanied by climatic changes triggered by the disruption of the monsoonal circulation (Parrish 1992), and directly implied a significant reorganization of water movements in the oceans, with the appearance of new barriers and the disappearance of previous ones. These facts should be taken into account when analyzing the distribution of marine benthic organisms.

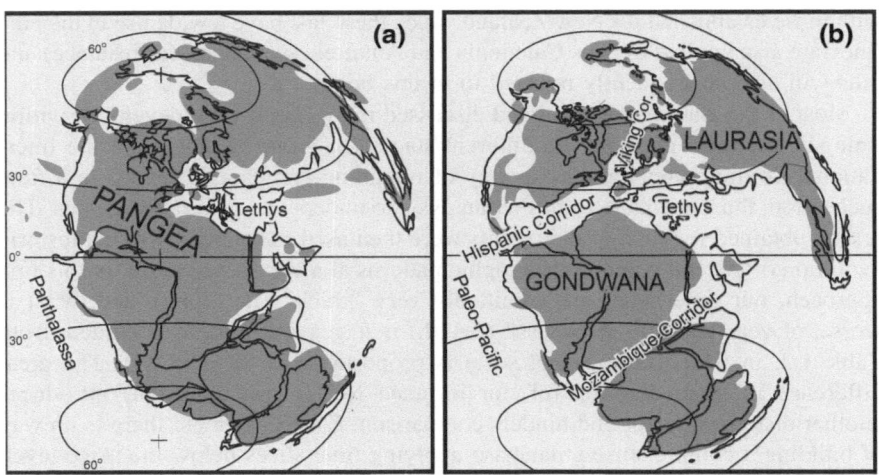

Fig. 1.2 Global paleogeography at the beginning (**a**, Triassic) and end (**b**, Late Jurassic) of the period of time considered in this book, compiled from various sources. Note that between the two moments the fragmentation of the Pangea supercontinent started, with the development of several marine corridors, among them the Hispanic Corridor, which connected the western Tethys with the Panthalassa (or Paleo-Pacific) and then originated the North Atlantic. These drastic changes affected oceanic circulation, climate, and the distribution of marine benthonic organisms (paleogeographic maps compiled from several sources, base map from Smith et al. 1981)

1.5 Paleoclimates and Water Temperatures

Most authors agree that the Jurassic was a period characterized by temperature gradients less evident than at the present (Hallam 1975, 1994), and although possible glaciations during the Jurassic were investigated, evidence is not convincing. A lot of geologic information backed by paleontologic knowledge of terrestrial faunas and floras supports this general statement. Research on this aspect is heavily dependent on the development of models (Parrish 1992).

The paleoclimatic aspect which interests us most in view of the subject of this analysis is sea-surface water temperatures. Most of the previous research was developed on Northern Hemisphere data, based on diverse isotopic analyses, and extrapolated to the whole Earth. According to Golonka and Ford (2000), greenhouse conditions prevailed during the Sinemurian-Toarcian, with a warm, humid environment, and moderate temperatures into high latitudes with no evidence of significant continental glaciation. On several evidences, Price (1999) concluded that the extent of polar ice during the Mesozoic was probably only one-third the size of the present day. Kiessling and Scasso (1996) suggested that Antarctic surface waters may have been warmer in average than those in equivalent northern high latitudes, according to the distribution of Pantanelliidae radiolarians.

The recent application of new techniques (TX_{86}) to Callovian to Late Jurassic deposits from the Southern Ocean (Jenkyns et al. 2012) supports the existence of

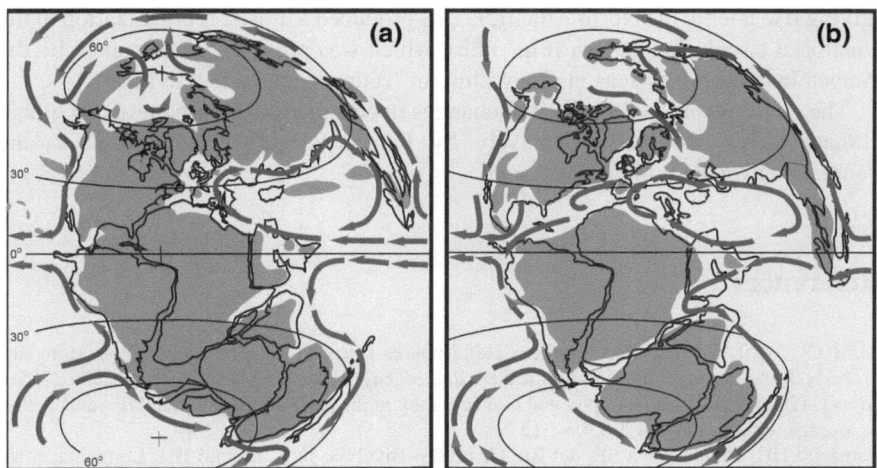

Fig. 1.3 Sketch of inferred global ocean surface circulation patterns for the Late Triassic-Early Jurassic (**a**) and the Late Jurassic (**b**), simplified from many sources. Only the inferred summer circulation pattern is shown for the western Tethys, a winter countercurrents pattern was strong during the Triassic but was not so clear after the opening of the Hispanic Corridor. Base map as in Fig. 1.2

relatively warm sea-surface conditions (26–30°) for that period. According to these studies, there was a general warming trend through the Late Jurassic, and the Callovian-Oxfordian boundary had slightly colder seawater temperatures (though never below 25°). These results strongly favor an equable tropical to subtropical environment up to the poles, contrary to earlier studies. It is now suggested that most paleotemperature studies based on belemnites, which give consistently lower temperatures, in fact reflect the water conditions below the termocline.

1.6 Paleocurrents

Paleoceanographic global models developed for the Late Triassic and Early Jurassic (Fig. 1.3a) propose that Panthalassa surface water circulation pattern appeared to be hemispherically symmetric, there was no circum-equatorial current, and circulation within the Tethys ocean had seasonal countercurrents (see revision in Parrish 1992; Arias 2006, 2008, and references therein). The annual surface circulation in Panthalassa had two main large subtropical gyres rotating anti-clockwise in the Southern Hemisphere and clockwise in the Northern Hemisphere. The circulation within the Tethys Ocean was quite different; in western Tethys the strong monsoonal regime produced an alternating change in water circulation directions between summer and winter.

The major paleogeographic global change produced during the Jurassic was the separation of Gondwana and Laurasia and the development of the Hispanic Corridor (giving rise later to the North Atlantic). This produced a drastic reorganization of the equatorial circulation pattern (Fig. 1.3b), which was previously interrupted by the Pangea landmass, and seasonality within the Tethys became less evident.

The consequences of these great changes in ocean circulation patterns on global climate were discussed extensively by Parrish (1992) with relation to the Panthalassa (or Paleo-Pacific) ocean.

References

Arias C (2006) Northern and Southern Hemispheres ostracod palaeobiogeography during the early Jurassic: possible migration routes. Palaeogeogr Palaeoclimat Palaeoecol 233:63–95

Arias C (2008) Palaeoceanography and biogeography in the early Jurassic Panthalassa and Tethys oceans. Gondwana Res 14:306–315

Camacho HH, Damborenea SE, del Río CJ (2008) Bivalvia. In: Camacho HH, Longobucco MI (eds) Los Invertebrados Fósiles. Tomo I. Fundación de Historia Natural Félix de Azara, Vazquez Mazzini Eds, Buenos Aires

Cecca F (2002) Palaeobiogeography of marine fossil invertebrates. Concepts and methods. Taylor & Francis, London

Cecca F (2009) La dimension biogéographique de l'évolution de la Vie. CR Palevol 8:119–132

Cecca F, Morrone JJ, Ebach MC (2011) Biogeographical convergence and time-Slicing. Concepts and methods in comparative biogeography. In: Upchurch P, McGowan AJ, Slater SC (eds) Palaeogeography and palaeobiogeography. Biodiversity in space and time. The Syst Assoc Spec vol 77. pp 1–11

Cooper RA (ed) (2004) The New Zealand geological timescale. Institute of Geological and Nuclear Sciences Monogr 22, Wellington

Crame JA (1986) Late Mesozoic bipolar bivalve faunas. Geol Mag 123:611–618

Crame JA (1987) Late Mesozoic bivalve biogeography of Antarctica. Proceeding 6th Gondwana Symposium, Columbus, Ohio, pp 93–102

Crame JA (1993) Bipolar molluscs and their evolutionary implications. J Biogeogr 20:145–161

Flessa KW, Jablonski D (1995) Biogeography of recent marine bivalve molluscs and its implications for paleobiogeography and the geography of extinction: a progress report. Hist Biol 10:25–47

Golonka J, Ford D (2000) Pangean (Late Carboniferous-Middle Jurassic) paleoenvironment and lithofacies. Palaeogeogr Palaeoclimatol Palaeoecol 161:1–34

Hallam A (1967) The bearing of certain palaeozoogeographic data on continental drift. Palaeogeogr Palaeoclimatol Palaeoecol 3:201–241

Hallam A (1969) Faunal realms and facies in the Jurassic. Palaeontology 12:1–18

Hallam A (1971) Provinciality in Jurassic faunas in relation to facies and palaeogeography. In: Middlemiss FA, Rawson PF, Newall G (eds.) Faunal provinces in space and time. Geol J Spec Issue 4:129-152

Hallam A (1975) Jurassic environments. Cambridge University Press, Cambridge

Hallam A (1977) Jurassic bivalve biogeography. Paleobiology 3:58–73

Hallam A (1994) Jurassic climates as inferred from the sedimentary and fossil record. In: Allen J, Hoskins B, Sellwood B, Spicer R, Valdes P (eds) Palaeoclimates and their modelling with special reference to the Mesozoic Era. Chapman & Hall, London

Hayami I (1961) On the Jurassic pelecypod faunas in Japan. J Fac Sci, UnivTokyo, Sect II Geol Mineral Geogr Geophys 13:243–343

Hayami I (1984) Jurassic marine bivalve faunas and biogeography in southeast asia. Geol Palaeontol Southeast Asia 25:229–237

Hayami I (1987) Geohistorical background of Wallace's line and Jurassic marine biogeography. In: Taira A, Tashiro M (eds) Historical biogeography and plate tectonic evolution of Japan and Eastern Asia. Terra, Tokyo

International Commission on Stratigraphy (2009) International stratigraphic chart

Jenkyns HC, Schouten-Huibers L, Schouten S, Sinninghe Damsté JS (2012) Warm Middle Jurassic-Early Cretaceous high-latitude sea-surface temperatures from the Southern Ocean. Climate Past 8:215–226

Kauffman EG (1973) Cretaceuos bivalvia. In: Hallam A (ed) Atlas of palaeobiogeography. Elsevier, Amsterdam

Kiessling W, Scasso R (1996) Ecological perspectives of Late Jurassic radiolarian faunas from the Antarctic Peninsula. In: Riccardi AC (ed) Advances in Jurassic research. Geo Res Forum 1-2:317-326

Lieberman BS (2000) Paleobiogeography. Using fossils to study global change, plate tectonics, and evolution. Topics in Geobiology 16, Kluwer Academic, Plenum Publishers, New York

Lieberman BS (2003) Paleobiogeography: the relevance of fossils to biogeography. Ann Rev Ecol Systemat 34:51–69

Lieberman BS (2008) Emerging synthesis between palaeobiogeography and macroevolutionary theory. Proc Roy Soc Victoria 120:51–57

Liu C (1995) Jurassic bivalve palaeobiogeography of the proto-atlantic and the application of multivariate analysis methods in palaeobiogeography. Beringeria 16:3–123

Morrone JJ (2009) Evolutionary biogeography: an integrative approach with case studies. Columbia University Press, New York

Niu Y, Jiang B, Huang H (2011) Triassic marine biogeography constrains the palaeogeographic reconstruction of Tibet and adjacent areas. Palaeogeogr Palaeoclimatol Palaeoecol 306:160–175

Parrish JT (1992) Jurassic climate and oceanography of the Pacific region. In: Westermann GEG (ed) The Jurassic of the circum-Pacific. IGCP Project 171, Cambridge University Press, Cambridge

Price GD (1999) The evidence and implications of polar ice during the Mesozoic. Earth Sci Rev 48:183–210

Rex MA, Crame JA, Stuart CT, Clarke A (2005) Large-scale biogeographic patterns in marine mollusks: a confluence of history and productivity? Ecology 86:2288–2297

Ros S, De Renzi M, Damborenea SE, Márquez-Aliaga A (2012) Early Triassic-Early Jurassic bivalve diversity dynamics. Treatise Online 39, Part N, Revised, vol 1. Chapter 25:1–19

Rosen BR (1992) Empiricism and the biogeographical black box: concepts and methods in marine palaeobiogeography. Palaeogeogr Palaeoclimatol Palaeoecol 92:171–205

Rosen BR (1995) Neobiogeography versus palaeobiogeography. In: Matteucci R, Carboni MG, Lee MST (eds) Studies on ecology and paleoecology of benthic communities. Boll Soc Paleontol Ital, Vol Esp 2:291-303

Sepkoski JJ jr (1981) A factor analytic description of the Phanerozoic marine fossil record. Paleobiology 7:36–53

Smith AG, Hurley AM, Briden JC (1981) Phanerozoic paleocontinental world maps. Cambridge University Press, Cambridge

Stanley SM (1970) Relation of shell form to life habits in the Bivalvia (Mollusca). Mem Geol Soc Am 125:1–296

Wallace AR (1855) On the Law which has regulated the introduction of new species. Ann Mag Nat Hist (2nd Ser) 16:184–196

Zinsmeister WJ (1979) Biogeographic significance of the late Mesozoic and early Tertiary molluscan faunas of Seymour Island (Antarctic Peninsula) to the final breakup of Gondwanaland. In: Gray J, Boucot AJ (eds) Historical biogeography, plate tectonics, and the changing environment. Oregon State University Press, Cambridge

Zinsmeister WJ (1982) Late Cretaceous-early Tertiary molluscan biogeography of the southern circum-Pacific. J Paleontol 56:84–102

Chapter 2
Techniques

Abstract Databases used for the analysis of past biotas should be as internally consistent as possible taking into account the incompleteness of the fossil record and the taxonomic distortions due to the history of their knowledge. A comprehensive and critically updated database of Southern Hemisphere bivalve occurrences through the Triassic and Jurassic was built. Most of paleobiogeographic analyses were performed within time slices to obtain comparable results in a time succession. Analytical methods were used for both (a) the analysis of latitudinal ranges along the South American paleo-coasts, and (b) the recognition of paleobiogeographic units for the Southern Hemisphere. a) The first approach to the study of species latitudinal ranges was cluster analysis, but this method, although useful, imposes a hierarchical structure on the data. Thus, to check for faunal changes along latitude, the distribution limits of species were explored using a technique similar to that considered for origination/extinction analysis, substituting first and last appearances by northernmost and southernmost geographical occurrences. Generalized linear models were used to look for changes on the proportional values of different species categories related to systematic and paleobiogeographic kinships. b) For the recognition of biochoremas, the incomplete and uneven nature of the data precludes the application of methods which may group areas according to the common absence of data, and we followed a traditional approach based on endemicity. In order to check the biogeographic structures without assuming a hierarchical or gradational arrangement, a Bootstrapped Spanning Network was calculated.

2.1 The Data

Occurrences of Triassic and Jurassic bivalve species were compiled from various published sources as well as the authors' own data, and plotted stage by stage from Induan (Early Triassic) to Berriasian (Early Cretaceous). Ros' (2009; also

S. E. Damborenea et al., *Southern Hemisphere Palaeobiogeography of Triassic-Jurassic Marine Bivalves*, SpringerBriefs Seaways and Landbridges: Southern Hemisphere Biogeographic Connections Through Time, DOI: 10.1007/978-94-007-5098-2_2, © The Author(s) 2013

Ros et al. 2012) updated database was used for Triassic occurrences. The study area is restricted to the Paleo-Southern Hemisphere, but the initial database was compiled on a global scale, not only to provide the necessary framework for the detailed analysis of southern regions, but also to adequately recognize patterns of general distribution and endemism. The species distribution data compilation was systematically and stratigraphically updated as far as possible and dubious records were excluded. The most serious problems related to such global databases are the incompleteness of the fossil record and the taxonomic distortions introduced by different authors working in different areas at different times. The first problem has no immediate solution, while the second can be somewhat reduced by critical evaluation of the data. This cannot always be done, for several reasons, but internal consistency was sought whenever possible, with careful reappraisal of both taxonomy and age of the records taken from the previous literature. In order to obtain a sound foundation for biogeographic considerations, only species personally examined or adequately described and figured were included; uncritical listing of taxa from sources lacking illustrations was avoided.

Presence-absence data were used throughout, since reliable quantitative records are only available for a small fraction of the occurrences.

Theoretically, the species is the most objective of taxonomic units; however, when species lists are compiled from studies made by various authors and at different times they become intensely subjective to the point that compilations at the generic or familial levels are preferred for global analysis (Stehli et al. 1967). Furthermore, the use of genus-group taxa increases the consistency of the database, as generic and sub-generic concepts have more consensus than species among different authors. In this book, genera and subgenera are used for the larger scales analyses (hemispheric and global), while species are preferred for performing regional analysis within the area studied by the authors, where first-hand knowledge facilitates identification and consistency.

Although it is evident that there are serious gaps in our knowledge of Triassic and Jurassic faunas from certain regions and for some bivalve groups, which have not been systematically reviewed or updated yet, it is believed that the data are comprehensive enough for the general purpose of this study.

In order to rationalize the study, the biogeographic affinities of the species (according to the categories discussed in Sect. 3.3) were recorded as well. Since the assignment of species to a definite type is based on its known distribution and relative abundance, and this knowledge is constantly being improved, this task proved difficult in some instances but relatively straightforward in others (such as the pectinaceans, see Hayami 1989; Damborenea 1993). Though many of the species have local distributions restricted to South America, they may have strong relations with other species or genera belonging to the paleobiogeographic affinities categories recognized here, and these were listed and used in the analysis.

Relative abundance was regarded as an important factor too; sporadic occurrences outside the main area of distribution are to be expected and, if adequately recognized as such, should not obscure the picture.

2.2   Quantification: A Difficult Approach

The application of any analytical technique in biogeography implies the acceptance of a defined biogeographic theory; biogeographic patterns are the result of ecological and historical processes acting on them, and if one accepts that such patterns actually do exist, then some analytical approach is required (Lieberman 2000). Quantitative analysis are claimed to be necessary in order to make biogeography a more rigorous science; these kinds of approaches should act as arbiters to choose between competing hypotheses. Nevertheless, these analytical techniques on their own do not guarantee valid results (Lieberman 2000), and the quality of the outcome largely depends of the quality of the available data.

The basic unit of any biogeographic study is the geographic range of a species, but its characterization is complicated by problems of defining and mapping them; what we usually see as maps of geographic ranges are simplifications of complex historic and ecologic patterns (Brown et al. 1996). Species ranges are usually mapped as irregular continuous areas, although in some cases they are presented as more precise "dot maps" that plot each location where a species has been recorded; most published range maps attempt to define the historical range of a species, encompassing localities where a species has regularly occurred in the past and recently colonized areas (Brown et al. 1996). To all these problems, common in the analysis of modern biota, the imperfection of the fossil record and the time averaging are added on paleobiogeographic analyses.

We use reconstructions of Mesozoic paleogeography to discuss our results. As pointed out by Smith (2011), although we now know a great deal more than 50 years ago, available Mesozoic paleogeographic maps are still rudimentary, showing only where continents and oceans were in the past, without bathymetry data, for instance. For the general purposes of this book they are nevertheless adequate.

Like any other model, a biogeographic model is a simplification of reality, and hence, as pointed out above, considered geographic ranges are simplified from actual geographic ranges; the kind of simplification applied will depend on the aim of the study.

In Sect. 4.2, an analysis of bivalve distributions along the west paleo-coast of South America is addressed; this coast presents a north–south orientation nowadays and it was similarly disposed during the Jurassic. This peculiarity allows for an analysis of latitudinal gradients on bivalve distribution. These gradients are associated with climatic gradients, imposing ecological restrictions to species distribution. A total of 13 areas were defined between parallels 20° and 46° S, each with a latitudinal range of 2°, as considered by Damborenea (1996). The presence of a species in one locality was computed as a presence for the whole area. During this analysis, discontinuity of the outcrops and differences in preservation probability were interpreted to be biasing a somewhat continuous distribution (Fig. 2.1), and so species ranges were interpolated between two localities with occurrence

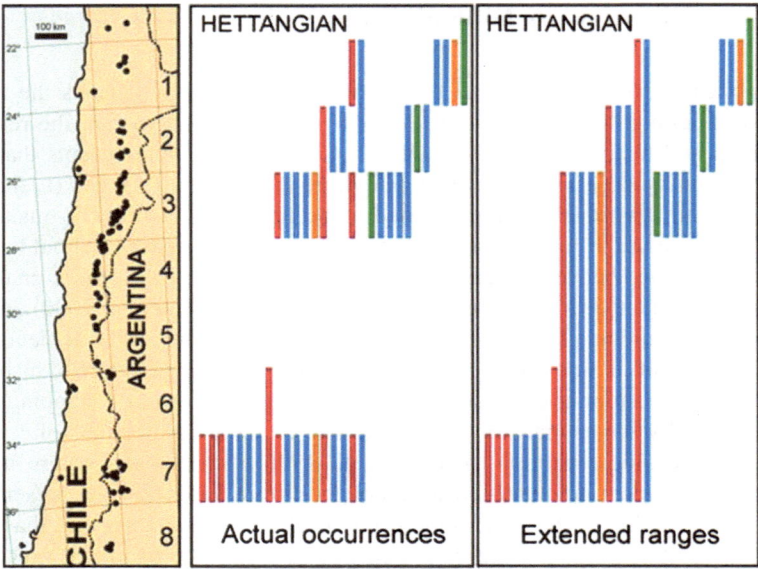

Fig. 2.1 Latitudinal range data exemplified with the Hettangian species data along a section of the eastern Paleo-Pacific coast. The actually observed occurrences (*left*) were used in most of the analyses performed, while the graphic presentation of data in Sect. 4.2.2 (Figs. 4.7, 4.9, 4.11 and 4.13) was simplified to show only the extended ranges (*right*)

(although on certain analyses, explained below, only the observed data were considered).

In Sect. 5.2, a hemispheric-scale analysis was carried out, and for this the units considered were well defined and bounded regions established a priori based on the current literature on the topic. Presence or absence of taxa (genera group taxa in this case) was computed for each region as a whole, disregarding the internal structure of distribution. The detail lost with this approach is not relevant at the scale considered for the purposes of this review.

Concerning time, most of the analyses performed were developed within time slices coinciding with the different stages (or group of stages), as a way of obtaining comparable results in a time succession. If stages are used, time slices are very unevenly "thick" (see Table 1.2, in which the vertical scale is proportional to time in Ma). One way to overcome in part the unevenness of the time duration of each slice is to group stages as done for some of the analyses (see Table 5.1), a method which also has some other practical advantages during the data gathering process.

On top of this, the time averaging within each of these time slices is considerable. The effects of time averaging are difficult to remove, especially in small-scale studies; as mentioned before, "instantaneous" geographic ranges cannot be obtained even for modern species, and this is yet less likely for fossil species. To deal with this, time bins were considered, and stratigraphic ranges are bounded by their upper and lower limits. The geographic ranges presented here include all the localities, where a species was present at any time of that time bin, in this case a stage; in some cases, the paucity of data forced the grouping of stages analyzed.

2.3   Analytic Methods

Phenetic approaches are common techniques in biogeographic analysis; in these methods information about distribution of taxa is entered into a data matrix for each of the regions or areas analyzed (Lieberman 2000). For this particular work only presence or absence of taxa (species or genera, depending on the analysis) was considered; abundance was indirectly considered solely when establishing biogeographic affinities.

All calculations were done using the different facilities provided by the softwares PAST (Hammer et al. 2001) and R (R Development Core Team 2008), as well as standard spreadsheet software.

Analytical methods further used to deal with both the latitudinal distribution and the biochoremas recognition are shortly described below.

2.3.1   Latitudinal Distributions

2.3.1.1   Cluster Analysis

The first explorative technique applied here is the hierarchical cluster analysis, for which a distance or similarity measure must be defined (Hammer and Harper 2006). Our main goal was to group together the localities according to their species content, so the Simpson's coefficient of similarity (Simpson 1943; see also Shi 1993) was used. This index is defined as the number of shared species between two localities divided by the number of species in the smaller sample. This index is totally insensitive to the size of the larger sample, which makes it suitable when sampling is considered to be incomplete (Shi 1993; Hammer and Harper 2006), as is the case for our database. The localities considered here were not equally treated in the literature, neither have the same abundance of fossils; hence, they cannot be considered as equally sampled, making the Simpson's coefficient the most adequate available index of similarity to use. Cluster analysis is an ordination method, grouping elements according to their similarity; clusters or groups have no statistical significance associated. A support value can be obtained for the nodes by simply resampling taxa (in this case species) and building a new dendrogram; the proportion of times the node appears on the dendrograms resulting from the resampled matrices is the support value for the node. Although the general grouping and disposition of the localities are evaluated on each analysis, special value is given to groups with similarity values of 0.50 or higher (i.e., 50 % of species shared or more) and to groups with support values of 0.50 or higher, as considered in other paleobiogeographic studies (Brayard et al. 2007; Dera et al. 2011).

On the analysis of latitudinal gradients the main interest focuses on distribution limits, so for species that appear at two distant localities it is usual to extend their ranges along the intermediate latitudes, and most data here are presented graphically in that way for clarity (Fig. 2.1). Nevertheless, for this analysis that methodology would result in a circular reasoning, since the latitudinal gradient would be analyzed presuming its existence; nearby localities would be similar because we assume they share species for being close to each other. To avoid this, cluster analysis was performed on the actually observed presence/absence data; this may produce some sensitivity to differences in knowledge between localities, but that is why we use Simpson's coefficient.

2.3.1.2 Distribution Limits of Species

To check for the faunal changes along a latitudinal gradient, we analyzed the distribution limits of the considered species through that gradient. Cluster analyses, although useful, are hierarchical ordination methods and hence they impose a hierarchical structure on the data, whether this exists or not. If a gradation among localities is to be expected, as happens in a latitudinal gradient, other independent approaches should be considered to check for it. A first graphic and very simple approach is to analyze the distribution limits of the considered species through that gradient. The methodology applied is similar to that considered for origination/ extinction analyses, counting the first and last appearance data (FAD and LAD respectively) on each stage (Hammer and Harper 2006), although in this case the stages are substituted by the latitudinal intervals, while the FADs and LADs are replaced by the northern distribution limit datum (NDL) and the southern distribution limit datum (SDL). If faunal turnover presents a gradational pattern, then high values of SDL and NDL are expected in all areas. On the other hand, sudden changes in faunal distribution will be recognized as peaks on the graphic; particularly significant will be the coincidence of peaks on both curves since they will show a major faunal turnover at that latitude (i.e., there will be a lot of species that appear only to the north and a lot that appear only to the south of that point). Peaks on only one curve indicate a reduction on general diversity on one direction (either north or south) and may be informative depending on the nature of data. This reduction could be spurious, if it only represents a sampling bias. For example, on the graphic for the Pliensbachian stage (Fig. 4.14), there is a high peak on the NDL curve between 24° and 26° S, but data for the areas between 20° and 24° S are scarce, and hence many of the considered species may have a broader range, extending northwards; something similar may be happening on the SDL peak between 40° and 42° S.

2.3.1.3 Generalized Linear Models

A second approach to check for gradational patterns is to look for changes on the proportional values of different species categories; data such as biogeographic affinities or systematic kinship (for instance, superfamilies) are good raw material for this kind of analysis. Generalized linear models (GLMs) are useful for data on proportions (Crawley 2007). The software R (R Development Core Team 2008) carries out a weighted regression, using the individual sample sizes as weights and the logit link function to ensure linearity (Crawley 2007). As a result a linear predictor is obtained together with its significance; the significance level used here was 0.05, but significance values between 0.05 and 0.10 were also considered as potentially explanatory. Positive linear predictors will imply positive associations between variables (i.e., an increment in the independent value, on this case latitude, is associated to an increment on the dependent value, on this case the proportion of species of the analyzed group). Negative linear predictors will imply the opposite trend, i.e., an increasing proportion of species of the group toward the north. This same analysis was applied to other groupings, such as superfamilies; on this last case the analyses were formed both on observed data and on extended range data. Localities poorly sampled may introduce noise instead to clear up a pattern, and so were removed from the analysis. When this happened it was made clear in the discussion.

2.3.2 Recognition of Biochoremas and Their Characterization

The recognition and characterization of "conventional" biogeographic units, i.e., based on area and hierarchy, was dismissed by some authors because "objective" criteria are lacking and methods used are not rigorous enough. One of the aims of biogeography is to infer historical connections among biotas or geographical areas on the basis of the distribution of organisms (historic biogeography). A variety of analytical methods were developed to compare the biota of different localities with this purpose (see a discussion and references in Posadas et al. 2006), one of the most popular being parsimony analysis of endemicity (PAE) developed by Rosen (1988).

The incomplete and uneven nature of our data precludes the application of methods which may group areas on the basis of the common absence of taxa. So at this stage we have mostly followed a traditional qualitative approach based on endemicity, combined with other features, such as diversity, to characterize biochoremas. In the same vein, we have consciously avoided the use of ranking names for the units so discussed (such as Realm, Province, etc.), and prefer to refer them simply as "biochoremas", sometimes specifying their relative ranking only.

Cluster analysis was applied to these data too, but only as a first approach since, as explained before, this kind of analysis imposes a hierarchical structure.

2.3.2.1 Bootstrapped Spanning Network

In order to check biogeographic structures between regions but without assuming either a hierarchical or a gradational structure, we calculated a Bootstrapped Spanning Network or BSN (Brayard et al. 2007) for the localities involved. The construction of a BSN is a straightforward procedure; starting from a matrix of observed occurrences a dissimilarity matrix was obtained, using as distance measure the complement value of Simpson's coefficient (1-S). From this second matrix a minimum spanning network (MSN) was obtained, by connecting localities from lower to higher dissimilarity value until all localities are directly or indirectly connected. Then, the original data matrix of observed occurrences was randomly resampled with replacement, and new dissimilarity matrix and MSN were calculated; this whole procedure was repeated 1000 times. For each edge on the original MSN a bootstrapped support value was obtained by simply calculating the proportion of times each edge was obtained on the original and resampled MSNs. Starting from the observed MSN, edges were removed (starting by those with the least support value) if and only if the network was still connected after its removal and the overall product of bootstrapped supports was increased; the resulting network is the bootstrapped spanning network (Brayard et al. 2007), obtained here for the more complete data sets.

BSN may be sensitive to the presence of sampled areas with low diversity. Lets consider as an extreme case a locality with only one taxon; if this taxon is endemic to that region, then the distance value of this area will be maximum (i.e. 1) with any other area. During the construction of the resampled MSN, if this taxon is sampled, all values lower than 1 (almost all edges) will be chosen, hence incrementing the support values of many edges that most probably should be considered as low supported. To avoid this, areas with less than 10 known taxa for the stage involved were removed from the analysis.

References

Brayard A, Escarguel G, Bucher H (2007) The biogeography of Early Triassic ammonoid faunas: clusters, gradients and networks. Geobios 40:749–765

Brown JH, Stevens GC, Kaufman DM (1996) The geographic range: size, shape, boundaries, and internal structure. Ann Rev Ecol Syst 27:597–623

Crawley MJ (2007) The R book. Wiley, NY

Damborenea SE (1993) Early Jurassic South American pectinaceans and circum-Pacific palaeobiogeography. Palaeogeogr Palaeoclimatol Palaeoecol 100:109–123

Damborenea SE (1996) Palaeobiogeography of Early Jurassic bivalves along the southeastern Pacific margin. 13° Congr Geol Argent, 3° Congr Explor Hidrocarb (Buenos Aires). Actas 5:151–167

Dera G, Neige P, Dommergues JL, Brayard A (2011) Ammonite paleobiogeography during the Pliensbachian–Toarcian crisis (Early Jurassic) reflecting paleoclimate, eustasy, and extinctions. Global Planet Change 78:92–105

Hammer Ø, Harper DAT (2006) Paleontological data analysis. Blackwell Publishing, Oxford

Hammer Ø, Harper DAT, Ryan PD (2001) PAST: Paleontological statistics software package for education and data analysis. Palaeontol Electron 4(1):9

Hayami I (1989) Outlook of the post-Paleozoic historical biogeography of Pectinids in the Western Pacific Region, The Univ Mus, Univ Tokyo. Nature Culture 1:3–25

Lieberman BS (2000) Paleobiogeography. Using Fossils to study global change, plate tectonics, and evolution. Topics in Geobiology 16, Kluwer Academic, Plenum Publishers, New York

Posadas P, Crisci JV, Katinas L (2006) Historical biogeography: a review of its basic concepts and critical issues. J Arid Environ 66:389–403

R Development Core Team (2008) R: a language and environment for statistical computing [Internet]. Vienna: R Foundation for Statistical Computing. http://www.R-project.org

Ros S (2009) Dinámica de la paleodiversidad de los Bivalvos del Triásico y Jurásico Inferior. PhD Thesis. Univ Valencia. Valencia. http://www.tesisenred.net/handle/10803/9952

Ros S, Márquez-Aliaga A, Damborenea SE (2012) Comprehensive database on Induan (Lower Triassic) to Sinemurian (Lower Jurassic) marine bivalve genera and their paleobiogeographic record. Paleontol Contrib, Univ Kansas (in press)

Rosen BR (1988) From fossils to earth history: applied historical biogeography. In: Myers AA, Giller PS (eds) Analytical biogeography: an integrated approach to the study of animal and plant distributions. Chapman and Hall, London

Shi GR (1993) Multivariate data analysis in palaeoecology and palaeobiogeography—A review. Palaeogeogr Palaeoclimatol Palaeocol 105:199–234

Simpson GG (1943) Mammals and the nature of continents. Am J Sci 241:1–31

Smith AG (2011) Uncertainties in Phanerozoic global continental reconstructions and their biogeographical implications. In: Upchurch P, McGowan AJ, Slater SC (eds) Palaeogeography and Palaeobiogeography. Biodiversity in Space and Time. Syst Assoc Spec 77:39–74

Stehli FG, McAlester AL, Helsley CE (1967) Taxonomic diversity in recent bivalves and some implications for geology. Geol Soc Am Bull 78:455–466

Chapter 3
A Bivalve Perspective

Abstract Mesozoic bivalves have been the subject of many paleobiogeographic studies, either with the aim of recognizing units, to argue about the proposal of opening of seaways and exotic terranes movements, or even to relate biogeography with extinction and evolution. With a few notable exceptions, Northern Hemisphere data were used and frequently conclusions extrapolated worldwide. In the analysis of bivalve geographic distribution, some special issues should be taken into account, such as larval type, mode of life, and tolerance to certain environmental factors, which are here briefly discussed for Southern Hemisphere bivalves. Special attention is paid to the proposed pseudoplanktonic habit as an aid to dispersal, to reef-building bivalves, and to those with special low-oxygen tolerance. For some of the various analyses performed, Triassic-Jurassic bivalve genera were classified according to their paleobiogeographic affinities in truly cosmopolitan, low-latitude (Tethyan), high-latitude (austral or bipolar), trans-temperate (Pacific), and endemic.

The study of past bivalve distributions has many applications, both geologic and biologic. Perhaps, the geologic ones have been explored more and are better known, but biologic implications are also diverse. For instance, it has been argued that geographic range in mollusks is significantly heritable (Jablonski and Hunt 2006) and thus it should be considered in evolutionary dynamic interpretations of these groups. In this chapter, we will try to discuss some general aspects which will be used in the analysis of bivalve distribution performed in the following chapters.

When dealing with fossil taxa, we have to bear in mind that we should refer all our observations to the past paleogeography, and frame them in the past climate and past ocean currents. Since all these features are mostly inferred, in many cases using the distribution of fossil taxa as evidence, it is indeed very difficult to avoid circular reasoning when arguing about these issues. Nevertheless, a general

S. E. Damborenea et al., *Southern Hemisphere Palaeobiogeography of* *Triassic-Jurassic Marine Bivalves*, SpringerBriefs Seaways and Landbridges: Southern Hemisphere Biogeographic Connections Through Time, DOI: 10.1007/978-94-007-5098-2_3, © The Author(s) 2013

discussion of each for the Triassic-Jurassic time interval and the Paleo-Southern Hemisphere is offered below.

3.1 Previous Research: A Northern Hemisphere Affair

Bivalves have provided data for paleobiogeographic studies almost since their earliest representatives (see Sánchez and Babin 2001; Cope 2002 on a summary of Ordovician bivalve paleobiogeography). The diverse Late Paleozoic bivalve faunas have particularly fueled interesting discussions related to regional and global paleogeography and paleoclimates (Runnegar and Newell 1971; Runnegar 1975; Dickins 1993; Shi and Grunt 2000, among many others).

The interest on Mesozoic bivalve distribution is also very old, and it was refreshed in the second half of the past century by new approaches based on abundant data, largely through the work of Anthony Hallam, Erle Kauffman and Graeme Stevens, who promoted a wide discussion on different (mainly geologic) related aspects. To make a complete account of previous research on the subject is well beyond the objectives of this book, and we will only mention a few significant contributions. Triassic and Jurassic bivalve geographic distributions were analyzed in many ways, not only from the descriptive point of view and the recognition of paleobiogeographic units (Hallam 1969, 1971, 1977; Fürsich and Sykes 1977; Kobayashi and Tamura 1983a, b; Silberling 1985; Hayami 1989, 1990; Tamura 1990; Liu et al. 1998; Grant-Mackie et al. 2000; Damborenea 2002; Shurygin 2005), but also as the main arguments to propose terrane movements (Newton 1983, 1987; Silberling et al. 1997; Aberhan 1998, 1999), migration routes (Marwick 1953), the opening of seaways (Damborenea and Manceñido 1979, 1988; Hallam 1983; Newton 1988; Liu 1995; Damborenea 2000; Aberhan 2001, 2002; Sha 2002), or even to relate biogeography with evolution (Ando 1987) and extinctions (Kiessling and Aberhan 2007), to explain the restriction to certain facies (Broglio Loriga and Neri 1976), or add to the geographic history of vast regions (Hayami 1961, 1984; Hallam 1967; Stevens 1967, 1977, 1980; Chen 1982; Hallam et al. 1986; Smith et al. 1990; Niu et al. 2011). They have been particularly instrumental to the discussion of general biogeographic issues such as bipolarity (Crame 1986, 1993, 1996; Damborenea 1993, 1998; Sha 1996) and latitudinal gradients (Damborenea 1996; Crame 2002; Niu et al. 2011) and the different factors governing distribution (Hallam 1981; Hayami 1987; McRoberts and Aberhan 1997).

Cretaceous bivalves and biogeography have also been the subject of many papers (for instance Kauffman 1973, 1975; Dhondt 1992, 1999), sometimes restricted to particular groups, such as rudists (Douvillé 1900; Coates 1973; Skelton and Wright 1987; Voigt et al. 1999; and many others) or specific biologic issues, such as evolutionary dynamics (Jablonski and Hunt 2006). There are also some recent contributions specifically related to the Southern Hemisphere: Aguirre-Urreta et al. 2008.

Cenozoic and living bivalves provide a wealth of information which has been used from the biogeographic viewpoint, either descriptively (e.g., Hall 1964; Emerson 1978; Zinsmeister 1979, 1982; Darragh 1985; Masse 1992) or methodologically (e.g., Jablonski and Valentine 1990), and the issue of latitudinal gradients in particular has been the subject of many discussions (e.g., Jablonski et al. 1999, 2000; Roy et al. 2000). In this case, the biologic significance of bivalve distribution is being addressed from many points of view, such as extinction (Flessa and Jablonski 1995).

3.2 Some General Issues

Despite the fact that most adult bivalves are not very mobile (except for some swimmers), many taxa have a remarkably wide biogeographic distribution. In living bivalves this is due to several biologic factors (already discussed by Kauffman 1975), as follows:

1. induced spawning when conditions are optimum for fertilization and larval survival,
2. a large egg yield,
3. a long-lived, planktonic larval stage, which may be extended if high-stress environments are imposed or suitable substrates are not encountered during settling,
4. a broad environmental tolerance in late larval and adult stages, including variable substrates and temporary, high-stress conditions,
5. effective adult mobility in some (swimmers),
6. "pseudoplanktonic" or epi-planktonic habits,
7. tolerance to extreme low-oxygen conditions.

On the whole, these factors are more difficult to evaluate when dealing with fossil taxa, but among them a planktotrophic larval stage can sometimes be directly observed in well-preserved material, while the swimming and pseudo-planktonic habits can be inferred from the morphology of the shell and/or taphonomic conditions, and tolerance to low-oxygen conditions is now supported by geochemic studies. For this reason, only these four factors will be discussed here in relation to Triassic and Jurassic taxa, with the addition of a few remarks about reef-forming bivalves during that time.

3.2.1 Larval Types and Dispersion

Bivalves have a wide variety of larval and developmental types, but most marine taxa have either planktotrophic or lecitotrophic-shelled larval stages, and these can sometimes be recognized in well-preserved specimens with preserved protoconch (see Jablonski and Lutz 1983 for a good review on the subject).

Knowledge on larval type distribution of living bivalves is scarce, among other things because they are difficult to identify since they usually have few distinctive features. For fossil species, information is still too incomplete. Unfortunately, present data on larval types of Mesozoic bivalves are limited to a small number of species, and there are even important groups for which direct evidence is not available yet. Sometimes the scanty information on living species is extrapolated and referred to the fossil taxa belonging to long-ranging families. However, this approach is still unsafe, since variability of this condition within groups is not sufficiently known. Moreover, Triassic and Jurassic marine bivalve faunas contain many groups with no living representatives, such as monotoids and inoceramoids. Within the first, an extended planktotrophic larval stage may have aided long-distance dispersal in halobiids (McRoberts 1997). Some information is also known about inoceramids: planktotrophic larvae in well-preserved Cretaceous material were described by Knight and Morris (1996) and Tanoue (2003); they also discussed the biogeographic significance for the dispersal of the group. The wide distribution of Triassic daonellids was attributed to long-term planktotrophic larvae (Campbell 1994; McRoberts 1997). For the Jurassic, larval shells are known in oxytomids and oysters (Palmer 1989), and bakevelliids (Malchus 2004).

Dispersal of the long-lived planktotrophic larvae is generally considered more effective than rafting of adults or egg masses (Kauffman 1975). However, this is still a controversial issue and cannot be applied without duly considering the particular conditions (e.g., O'Foighil 1989). Based on the study of two gastropod species, Johannesson (1988) argued that even if transport of benthic stages happens very rarely, this may be more influential than larval dispersal over long distances. Liu et al. (2007) studied the distribution of bivalve larval types along latitude in the east Pacific, and concluded that the average geographic dispersal levels of planktotrophic and non-planktotrophic larval types do not show great differences.

Pelagic planktotrophic larvae spend a large time (up to several weeks) in the water column before settling, and are thus mainly distributed by currents. During this time, food availability is one of the factors affecting their development and metamorphosis capacity. The long potential survival of holopelagic larval stage might allow for enough time to disperse even across large oceans (Scheltema 1977, 1988; Scheltema and Williams 1983) and although rare in living forms, extreme holopelagic adaptations were favored for several fossil taxa (Oschmann 1993), such as Triassic halobiids, for instance (McRoberts 1997).

Another aspect almost impossible to be considered for fossil species is the vertical distribution of planktotrophic larvae, which is normally stratified, and the associated depth is related to differences in water movement velocity. Bivalve larvae are capable of swimming and adjusting their vertical position in the water column (Raby et al. 1994).

In synthesis, although we recognize the great interest and potential of including larval types in the analysis of bivalve distribution, there is yet not enough information to do so without resorting to large-scale extrapolation. For these reasons no attempt will be done here to use inferred larval types in the analyses.

3.2.2 Swimming Habit

In the context of bivalve large-scale geographic dispersion, we are not interested in occasional swimmers but on truly nektoplanktonic bivalves, and these seem to have been particularly uncommon. The swimming habit is a late acquisition in bivalve evolution, but already in the Silurian some species had adaptive convergent features with Mesozoic Posidoniidae, which were regarded as nektoplanktonic (Kříž 1996).

However, at least a facultative swimming habit seems to have been comparatively widespread among Triassic bivalves, when members of the Posidoniidae (*Bositra* and others) and Halobiidae (*Halobia*, *Daonella*) were frequently mentioned as "pelagic" bivalves. The external morphology of some halobiid species could fit into a swimming mode of life (subcircular form, equivalve, and short hinge line), but the adductor muscle scar is small and falls just below the umbo, and the shells are too thin (Schatz 2005). All the discussion is far from settled, and it is possible that the different morphologies of members of the group indicate slightly different modes of life.

Several authors argued that at least some species of *Bositra* were nektonic (Jefferies and Minton 1965; Hayami 1969; Duff 1978). Also, *Posidonotis* can be included in this group, as it is frequently found in facial equivalents to the "Posidonienschiefer" of central Europe forming shell pavements. These bivalves are often called "paper-clams" (Wignall 1990) or "flat-clams" (see Aberhan and Pálfy 1996) and occur abundantly in dark shales. For most of them reclining and pseudoplanktonic habits were also proposed (see discussion below).

3.2.3 Pseudoplanktonic Habit

Exclusive pseudoplanktonic species are expected to be rare faunal elements, but preserved in a very wide range of environments and facies. Some epibyssate bivalves can have a facultative pseudoplanktonic habit if they fix to floating objects. There are several groups of bivalves which thrived during the Triassic and Jurassic and for which a mostly pseudoplanktonic habit has been proposed (Wignall and Simms 1990), though for few of them this is currently accepted as their main mode of life. These are thought to be mostly benthonic bivalves which occasionally could fix themselves to swimming organisms or floating objects, and so the role of this mode of life for geographic dispersal is not regarded as significant by some authors. On the other hand, in view of the extremely rapid dispersal (nearly instantaneous at geologic scale) experienced by some extant invasive bivalve species after accidental settlement of a few individuals in suitable habitats, rapid dispersal aided by pseudoplanktonic rafting cannot be dismissed. A few of the Triassic-Jurassic genera which could have facultative pseudoplanktonic species will be briefly mentioned below, and it is interesting to note that many of these species have a particularly wide geographic distribution:

Posidonotis Losacco: A pseudoplanktonic mode of life was proposed by Hillebrandt 1980, 1981) and extensively discussed by Aberhan and Pálfy (1996) on the basis of functional morphology, abundance, and facies distribution. These authors concluded that species of this genus had a mainly benthic mode of life.

Bositra de Gregorio: Several authors postulated the possibility of a pseudo-planktonic mode of life for this bivalve, on account of its preferred occurrence in dark, organic-rich mudstones (see for instance Stanley 1972), while others regarded it as nektoplanktonic. Etter (1996) re-evaluated the life habits of this bivalve and concluded that *B. buchi* (Roemer) was a truly benthic species and probably lived byssally attached to the substrate. An updated discussion on *Bositra* mode of life was provided by Caswell et al. (2009, see references therein).

Pseudomytiloides Cox: A mainly benthic mode of life was also favored for *P. dubius* (Sowerby) by Etter (1996), a species which is nevertheless known to occur as facultative pseudoplanktonic, as revealed by its occurrence in clusters arranged in lines (Caswell et al. 2009).

Halobia Bronn and *Daonella* Mojsisovics: Several modes of life were assigned to species of this group (Campbell 1994), although Schatz (2005) supported an epibenthic pleurothetic mode of life on soft substrate for *Daonella* species. However, other authors, such as McRoberts (1997) and Campbell (1994), suggested that a pseudoplanktonic mode of life cannot be ruled out, although Schatz (2005) argued that the morphology exhibited by halobiids does not support this hypothesis. McRoberts (1997) favored a pseudoplanktonic or epibenthic mode of life for this extremely diverse and short-lived bivalve group. He suggested that halobiids were probably not obligate pseudoplanktonic but, being opportunistic, could attach themselves to any available firm substrate (drifting or not).

3.2.4 Reef-Forming Bivalves

During severe biotic crises reef-forming organisms have particularly suffered, but the reef ecosystem has proven its resilience by being successively renovated after every extinction event. This is extensively documented in the interesting history of reef-building organisms through time, and in our particular time interval bivalves played a prominent role after the end Triassic extinction, which decimated corals, and just before the establishment of scleractinian coral reefs.

The classic deposits are collectively known as "*Lithiotis*" facies (named from the peculiar genus *Lithiotis* Gümbel) of the Calcari Grigi, and were first described from Pliensbachian-Toarcian beds in the southern Alps (Tausch 1890; Böhm 1906). The fossil associations are not diverse, and, besides particular gastropods and brachiopods, they include a variety of thick-shelled bivalves such as *Lithiotis*, *Cochlearites* Reis, *Lithioperna* Accorsi Benini, *Opisoma* Stoliczka, *Gervilleioperna* Krumbeck, *Mytiloperna* Ihering (in the sense of Accorsi Benini and Broglio Loriga 1982), *Pseudopachymytilus* Krumbeck, and some megalodontids, as *Pachymegalodon* Gümbel (Accorsi Benini 1979, 1981, 1985).

The geologic setting and both the geometry and the internal structure of the reef bodies were compared to the Cretaceous rudist limestones (Rey et al. 1990). The Alpine reefs are up to 5 m thick and tens of meters wide, and reef builders were *Lithiotis, Lithioperna,* and *Cochlearites* species. These exhibit a wide range of adaptations to this habitat and life habit, including very thick shells with discordant valves (one of them acting as a thin elastic operculum), allometric growth of the apical parts, and remarkable phenotypic variability (see discussion in Chinzei 1982 and Savazzi 1996). The possibility of symbiosis with green algae was proposed (Accorsi Benini and Broglio Loriga 1977) and later challenged on the basis of the analysis of growth increments (Accorsi Benini 1985), the large size being attributed to long life span (up to 30 years) at a normal growth rate.

Of special interest in the context of this book is the geographic distribution of the reef builders *Lithiotis* and *Cochlearites* (see maps in Broglio Loriga and Neri 1976; Nauss and Smith 1988; Krobicki and Golonka 2009), which were confirmed from Italy, Austria, Slovenia, Croatia, Albania, Morocco, Spain, Oman, Arab Emirates, Iran, Timor Indonesia, USA (where *Lithiotis* was called *Plicatostylus* Lupher and Packard), Peru, and northern Chile, and also reported from France and the Himalayas. This is a characteristic low-latitude, southern Tethyan distribution (see Sect. 3.3), and can be compared in latitudinal extension to the distribution of modern coral reefs. The *Lithiotis* bioherms and biostromes extended mainly along the southern margin of Tethys and reached both South and North America, but never extended to high paleolatitudes.

This particular bivalve association is the best Tethyan marker for Pliensbachian/Toarcian times and will be mentioned in several places of this book. It helps to define the boundaries of the Tethyan biochorema, and has also been used in the discussion of the establishment of the Hispanic Corridor.

3.2.5 Low-Oxygen Tolerance

Some Mesozoic bivalves are usually found associated to oxygen-poor sedimentary environments, their abundance in such settings being probably related to an opportunistic behavior (Etter 1996). These bivalves are collectively and informally called "paper pectens" or "flat clams" due to their usually very thin shell. The better studied are those from the lower Jurassic Posidonienschiefer in Europe. Data of the detailed relationship of such bivalves with true anoxic events are just being developed, and show that each species may have slightly different occurrence patterns. For instance, in Yorkshire *Bositra radiata* occurs at the beginning and the end of the main carbon-positive excursion but it is absent during the most severe

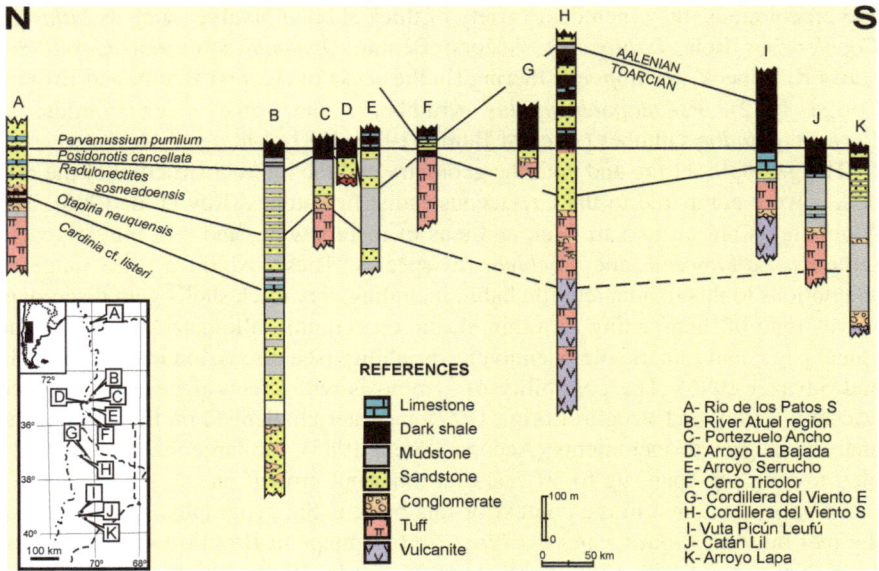

Fig. 3.1 Geographic distribution of *Posidonotis cancellata* (Leanza) in the Neuquén basin, Argentina. Inset map shows the location of the logged sections, the sketches are leveled to the top of the *Posidonotis*-bearing beds. Local bivalve biozonation indicated between sections *A* and *B*. See detail of the Arroyo Lapa section (*K, extreme right*) in Fig. 3.2

conditions during the Toarcian Oceanic Anoxic Event (TOAE), while *Pseudomytiloides dubius* is the only benthic taxon then present (Caswell et al. 2009).

In the western Americas, species of *Posidonotis* form shell pavements (see Fig. 4.4) in dysaerobic environments (Damborenea 1989; Aberhan and Pálfy 1996), and were qualified as low-oxygen tolerant by these last authors. *Posidonotis cancellata* (Leanza) has a wide geographic distribution in the Neuquén basin (Argentina) (Fig. 3.1).

A very detailed analysis of its distribution around the TOAE at the Arroyo Lapa section (Al-Suwaidi et al. 2010) shows that *Posidonotis* became abundant, forming dense monospecific shell pavements, just before the negative carbon isotopic excursion, but was not present either during or after the harshest conditions (Fig. 3.2).

McRoberts (1997) discussed the adaptations that enabled *Halobia* to live in poor-oxygen environments but concluded that there is only circumstantial evidence for a chemosymbiotic association with sulfur-reducing bacteria.

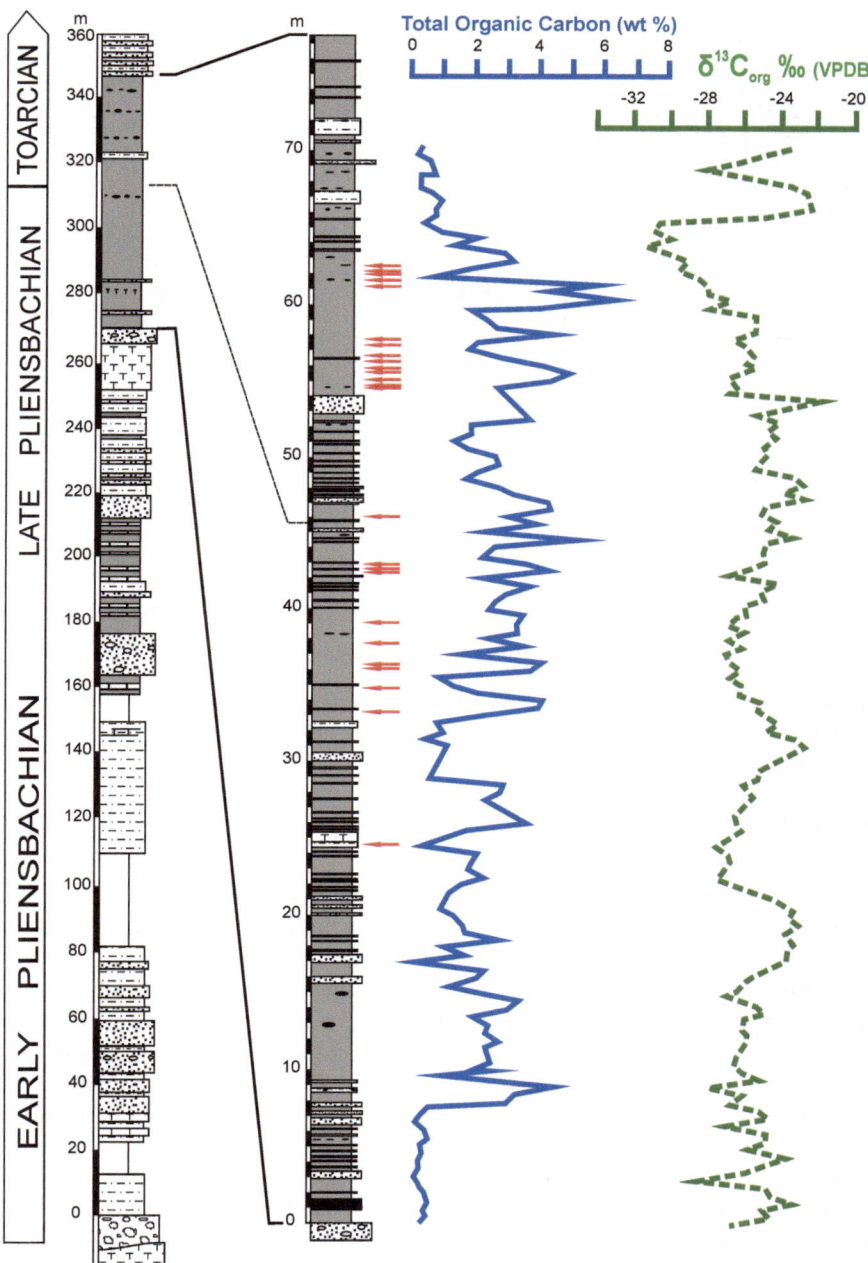

Fig. 3.2 Arroyo Lapa section, Neuquén basin, Argentina. General section to the *left*, detailed section of the Pliensbachian-Toarcian boundary beds to the *right*. Beds with *Posidonotis cancellata* (Leanza) shell pavements (*arrows*) related to the total organic carbon and $\delta^{13}C_{org}$. Chemostratigraphic data from Al-Suwaidi et al. (2010)

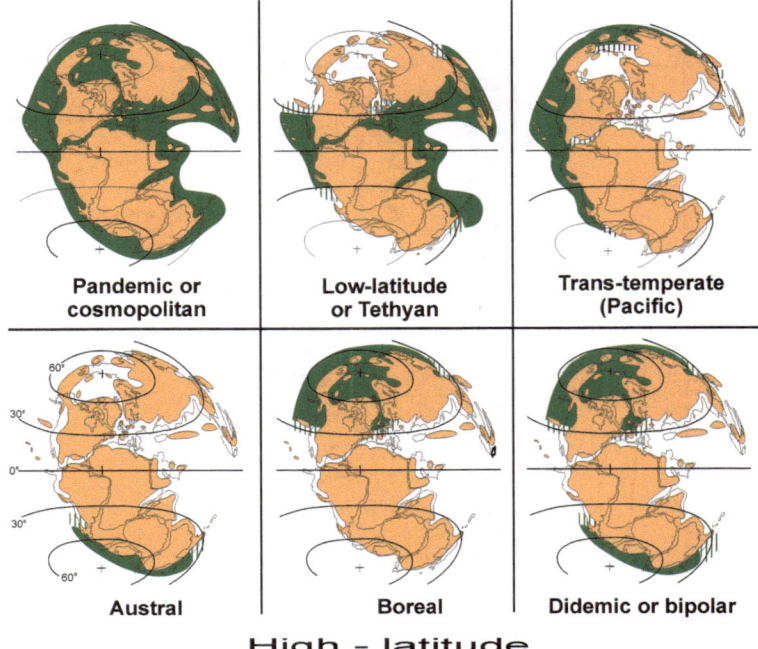

Fig. 3.3 General patterns of paleobiogeographic distribution recognized here for Triassic and Jurassic marine bivalves, plotted on a Pliensbachian-Toarcian paleogeographic map based on Smith and Briden (1977) and Scotese (1997)

3.3 Paleobiogeographic Affinities

For some of the various analyses exposed later in this book, bivalve genera were classified according to their paleobiogeographic affinities, following Kauffman (1973); Stevens (1980) and Damborenea (1993), as follows (Fig. 3.3):

Pandemic (cosmopolitan): A large number of genera are widespread bivalves, truly cosmopolitan forms, such as *Nuculana, Palaeoneilo, Mytilus, Pseudolimea, Daonella, Halobia, Entolium, Camptonectes, Eopecten, Bositra, Meleagrinella, Oxytoma, Neoschizodus, Trigonia, Septocardia, Pleuromya, Pholadomya*, etc. Several species of these genera have surprisingly wide geographic distributions. On the other hand, it is interesting to note that many genera traditionally referred to as "cosmopolitan" have not been reported from the Southern Hemisphere during the time interval considered. These include, for instance, *Atrina, Cirtopinna, Ctenoides, Limopsis, Linotrigonia, Malletia, Martesia, Megalodon, Mytiloides, Palaeomya, Paratancredia, Protocardia, Septifer*. Other so-called "cosmopolitan" genera are present only in the northern regions of the Southern Hemisphere but are not known from the southernmost areas. Examples are *Hippopodium, Neomegalodon, Pinguiastarte,* and *Rollieria*. All genera mentioned

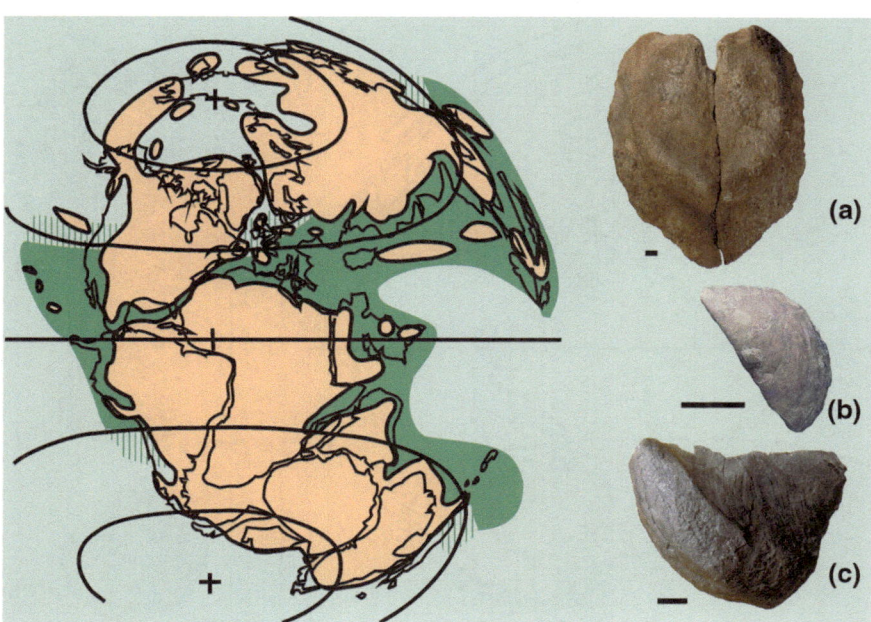

Fig. 3.4 General map of low-latitude distribution pattern and some examples of Tethyan bivalve genera: **a** *Opisoma*, **b** *Lycettia*, **c** *Gervillaria*. Specimens illustrated are: **a** *Opisoma* sp., MLP 18460, early Toarcian, Argentina; **b** *Lycettia* cf. *lunularis* (Lycett), MLP 19091, early Bajocian, Argentina; **c** *Gervillaria pallas* (Leanza), MLP 19079, Pliensbachian, Argentina. Scale bars: 10 mm. MLP: La Plata Natural History Museum

with such distributions were excluded from the list of "cosmopolitan" taxa for the analysis.

Low-latitude or Tethyan: Genera restricted to low paleolatitudes (Fig. 3.4). During the time involved the Tethyan sea extended along the low paleolatitudes. Tethyan bivalve faunas were characterized by their high diversity and the abundance of large, thick-shelled forms. Some examples are *Rhaetavicula, Curionia, Trichites, Spondylopecten, Radulopecten, Lycettia* (Fig. 3.4b), *Anningella, Caenodiotis, Krumbeckia, Wallowaconcha*, nearly all megalodontids, many arcoids and pterioids (Fig. 3.4c), and those restricted to *Lithiotis* reef facies as *Cochlearites, Lithioperna, Opisoma* (Fig. 3.4a), *Gervilleioperna, Pseudopachymytilus,* and *Pachymegalodon*.

High-latitude: Several Triassic and Jurassic bivalve genera show a geographic distribution restricted to areas that, according to most paleogeographic reconstructions, were at high latitudes during those times. As could be expected most of these high-latitude taxa were also restricted to the Panthalassa or Paleo-Pacific, the only ocean displaying the whole latitudinal range at that time. These genera were previously thought to be either East Asian or Southwest Pacific endemics, but now the following patterns can be recognized (Damborenea 1993):

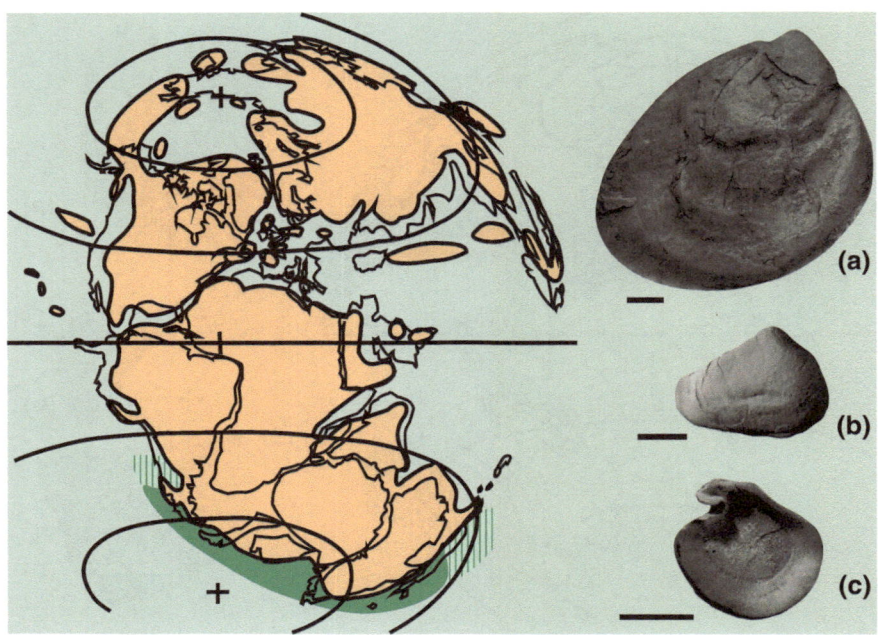

Fig. 3.5 General map of austral distribution pattern and some examples of Maorian bivalve genera: **a** *Oretia*, **b** *Pseudaucella*, **c** *Malayomaorica*. Specimens illustrated are: **a** *Oretia coxi* Marwick, NZGS-TM 2283 (holotype), Oretian, New Zealand (reproduced from Speden and Keyes 1981, pl. 9, Fig. 23); **b** *Pseudaucella marshalli* (Trechmann), OU 3501, Ururoan, New Zealand; **c** *Malayomaorica malayomaorica* (Krumbeck), NZGS-TM 5783, Heterian, New Zealand (reproduced from Speden and Keyes 1981, p. 14, Fig. 21). Scale bars: 10 mm. OU: Otago University. NZGS: New Zealand Geological Survey

1. *Austral* (= Maorian or paleoaustral). Only very few genera were widespread but restricted to high latitudes in the Austral regions (Fig. 3.5), as *Oretia*, *Malayomaorica*, *Pseudaucella*. Besides, there are a lot of genera which were endemic to particular austral regions, notably New Zealand and/or New Caledonia.
2. *Boreal*. No genera with such distribution reached the Paleo-Southern Hemisphere, and thus will not be discussed in detail here, but there were some bivalve genera restricted in distribution to boreal regions, as *Ochotochlamys* and *Amuropecten*.
3. *Didemic, antitropical, or bipolar* (Fig. 3.6), restricted to high latitudes and present in both hemispheres, being absent from the low-latitude intervening areas (Crame 1993; Damborenea 1993; Sha 1996). Examples are *Aparimella*, *Maoritrigonia*, *Minetrigonia*, *Ochotomya*, *Triaphorus*, *Asoella*, *Kalentera* (Fig. 3.6c), *Retroceramus* (Fig. 3.6a). Also, various pectinacean taxa, previously thought to be restricted and characteristic of Boreal regions, have been found in southern South America (Damborenea 1993) and their distribution was

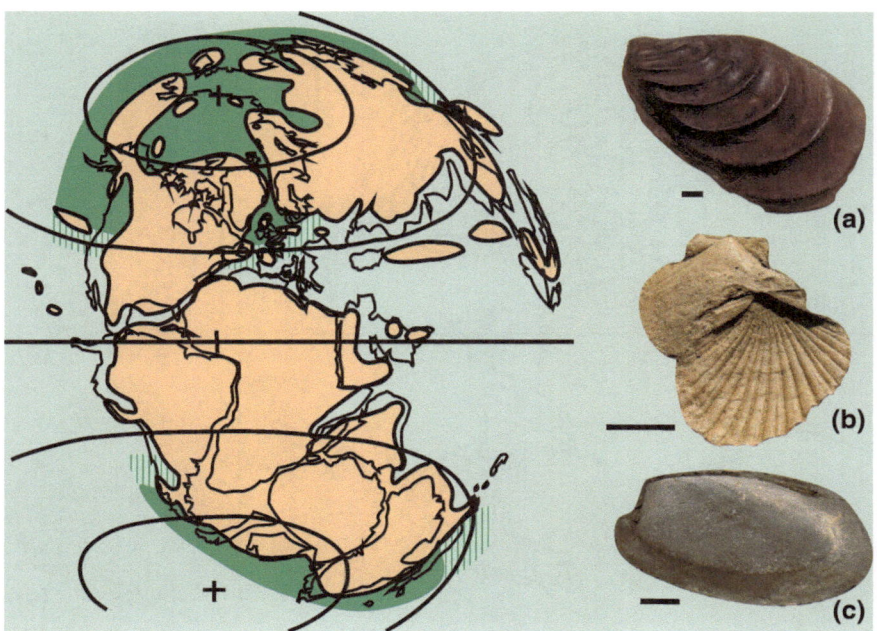

Fig. 3.6 General map of bipolar distribution pattern and some examples of bipolar bivalve genera: **a** *Retroceramus*, **b** *Kolymonectes*, **c** *Kalentera*. Specimens illustrated are: **a** *Retroceramus stehni* Damborenea, MLP 14672 (holotype), Early Callovian, Argentina; **b** *Kolymonectes weaveri* Damborenea, Late Pliensbachian, field photograph, Chubut, Argentina; **c** *Kalentera riccardii* Damborenea, MLP 24308, Pliensbachian, Argentina. Scale bars: 10 mm. MLP: La Plata Natural History Museum

in fact bipolar instead (Fig. 6.1), for instance *Palmoxytoma, Kolymonectes* (Fig. 3.6b), *Radulonectites, Agerchlamys, Arctotis*. This particular distribution pattern and representative genera will be discussed more extensively on Sect. 6.1 in relation to bipolarity.

Trans-temperate (Kauffman 1973): this category includes taxa common in temperate regions, but with a distribution not latitudinally limited, and at the same time not pandemic. For instance, bivalves with the so-called "Pacific" distribution (Fig. 3.7), i.e., along the margins of the Paleo-Pacific ocean both in low and medium (rarely also high) paleolatitudes, only occasionally present in other areas, such as *Otapiria* (Fig. 3.7a), *Weyla* (Fig. 3.7c), *Lywea, Posidonotis* (Fig. 3.7b), *Cardinioides,* and some trigonioidean genera, at least during part of their time range.

Endemic: These are the key elements to recognize and characterize biogeographic units. Only true endemics (from each region considered here compared at the global scale) were counted as such (Fig. 3.8). Endemism was high in various Southern Hemisphere areas in the Late Triassic, almost disappeared after the Triassic/Jurassic extinction, and was again significant for late Early Jurassic, to

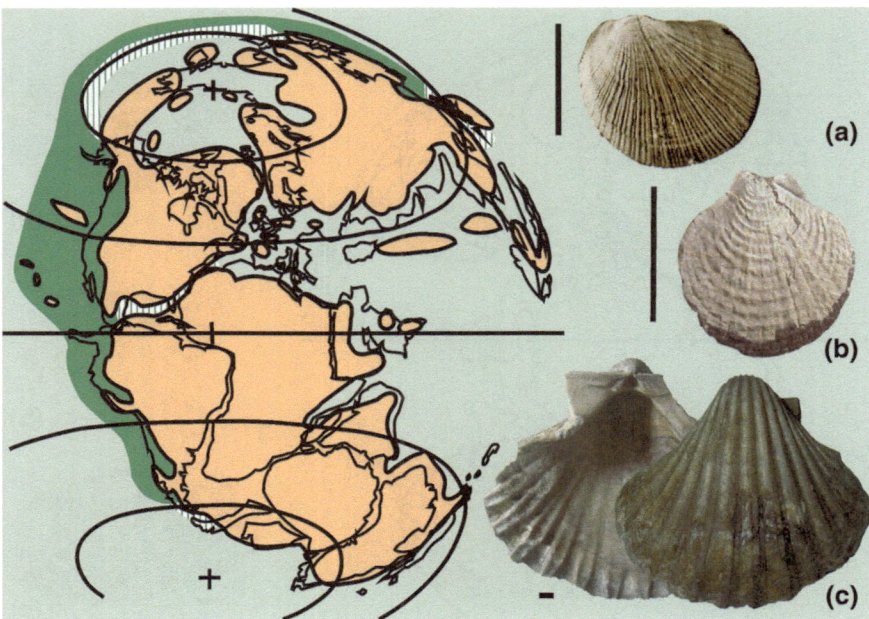

Fig. 3.7 General map of trans-temperate distribution pattern and some examples of Pacific bivalve genera: **a** *Otapiria*, **b** *Posidonotis*, **c** *Weyla*. Specimens illustrated are: **a** *Otapiria neuquensis* Damborenea, MLP 16480 (holotype), Pliensbachian, Argentina; **b** *Posidonotis cancellata* (Leanza), Early Toarcian, field photograph, Argentina; **c** *Weyla angustecostata* (Philippi), MLP 19076, Early Toarcian, Argentina. Scale bars: 10 mm. MLP: La Plata Natural History Museum

decline toward the end of the Jurassic. There are many genera which were endemic to New Zealand and/or New Caledonia, for instance *Etalia, Marwickiella, Agonisca, Caledogonia, Hokonuia* (Fig. 3.8c)*, Praegonia, Oretia, Ouamouia* in the Triassic; *Notoastarte, Haastina, Kanakimya, Moewakamya, Austrocardilanx* in the Jurassic. *Isopristes, Perugonia,* and *Schizocardita* (Fig. 3.8a) were endemic to the central Andean region in South America during the late Triassic, while in the Jurassic *Quadratojaworskiella, Gervilletia, Gervilleiognoma, Neuquenitrigonia, Andivaugonia* (Fig. 3.8b), *Eoanditrigonia, Anditrigonia,* and perhaps a few others were characteristic of that region. During the late Triassic some peculiar genera were exclusively known from Australia-New Guinea, as *Gervillancea, Guineana, Somareoides* (Fig. 3.8d), and *Krumbeckiella*. Other set of genera characterized the late Triassic bivalve faunas from Iran, such as *Antiquicorbula, Healeya, Modesticoncha, Primahinnites, Triasoperna,* and *Umbrostrea*.

Only pandemic and endemic taxa can be objectively defined, reference of some genera to either of the other categories may involve necessarily a debate. This is not a serious drawback in this context, however, since these last categories are only accessory elements in this study. From the nearly 500 bivalve genera known from

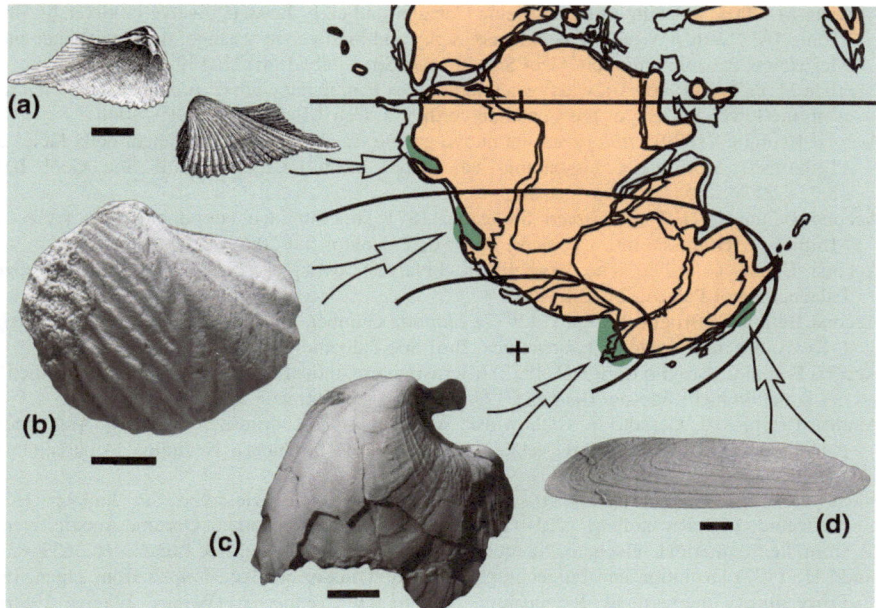

Fig. 3.8 Some examples of endemic bivalve genera from different regions and ages: **a** *Schizocardita,* **b** *Andivaugonia,* **c** *Hokonuia,* **d** *Somareioides.* Specimens illustrated are: **a** *Schizocardita cristata* Körner, Norian, Perú (reproduced from Körner 1937, Fig. 4); **b** *Andivaugonia radixscripta* (Lambert), MLP 6710, Bajocian, Argentina; **c** *Hokonuia limaeformis* (Trechmann), OU 17487, Otamitan, New Zealand; **d** *Somareioides hastatus* (Skwarko), Norian, New Guinea (reproduced from Skwarko 1983, p. 1, Fig. 9). Scale bars: 10 mm. MLP: La Plata Natural History Museum; OU: Otago University

the Jurassic, 52 % can be regarded as low latitude, 21 % as high latitude, 6 % as trans temperate, and 21 % as pandemic (Damborenea 2002, p. 53). It is interesting to note here that for the whole Jurassic, and despite the paleogeographic asymmetries mentioned above, endemic genera are proportionately equally frequent in high-latitude and in low-latitude regions.

References

Aberhan M (1998) Paleobiogeographic patterns of pectinoid Bivalves and the early Jurassic tectonic evolution of western Canadian terranes. Palaios 13:129–148

Aberhan M (1999) Terrane history of the Canadian Cordillera: estimating amounts of latitudinal displacement and rotation of Wrangellia and Stikinia. Geol Mag 136(5):481–492

Aberhan M (2001) Bivalve palaeobiogeography and the hispanic corridor: time of opening and effectiveness of a proto-Atlantic seaway. Palaeogeogr Palaeoclimatol Palaeoecol 165:375–394

Aberhan M (2002) Opening of the Hispanic Corridor and early Jurassic bivalve biodiversity. In: Crame JA, Owen AW (eds) Paleobiogeography and biodiversity change: the Ordovician and Mesozoic-Cenozoic radiation. Geol Soc London Spec Publ 194:127–139

Aberhan M, Pálfy J (1996) A low oxygen tolerant East Pacific flat clam (*Posidonotis semiplicata*) from the lower Jurassic of the Canadian cordillera. Can J Earth Sci 33:993–1006

Accorsi Benini C (1979) *Lithioperna*, un nuovo genere fra i grandi lamellibranchi della facies a "Lithiotis". Morfologia, tassonomia ed analisi morfofunzionale. Boll Soc Geol Ital 18:221–257

Accorsi Benini C (1981) *Opisoma* Stoliczka, 1871 lamellibranco eterodonte della facies a "Lithiotis" (Giurassic inf., Liassico). Boll Soc Paleontol Ital 20:197–228

Accorsi Benini C (1985) The large liassic bivalves: symbiosis or longevity. Palaeogeogr Palaeoclimatol Palaeoecol 52:21–33

Accorsi Benini C, Broglio Loriga C (1977) *Lithiotis* Gümbel, 1871 e *Cochlearites* Reis, 1903. I. Revisione morfologica e tassonomica. Boll Soc Paleontol Ital 16:15–60

Accorsi Benini C, Broglio Loriga C (1982) Microstructure, modalità di accrescimento e priodicità nei lamellibranchi liasssici (Facies a "*Lithiotis*"). Geol Romana 21:795–823

Aguirre-Urreta MB, Casadío S, Cichowolski M, Lazo DG, Rodríguez DL (2008) Afinidades paleobiogeográficas de los invertebrados cretácicos de la Cuenca Neuquina. Ameghiniana 45:593–613

Al-Suwaidi AH, Angelozzi GN, Baudin F, Damborenea SE, Hesselbo SP, Jenkyns HC, Manceñido MO, Riccardi AC (2010) First record of the Early Toarcian Oceanic Anoxic Event from the Southern Hemisphere, Neuquén Basin, Argentina. J Geol Soc London 167:633–636

Ando H (1987) Evolution and biogeography of Late Triassic bivalve *Monotis* from Japan. In: Proceedings International Symposium on Shallow Tethys 2 (Wagga Wagga 1986), pp 233–246

Böhm G (1906) Zur Stellung von Lithiotis. Centralbl Mineral Geol Paläontol 1906:161–167

Broglio Loriga C, Neri C (1976) Aspetti paleobiologici e palaeogeografici della facies a "*Lithiotis*" (Giurese inf.). Riv Ital Paleontol e Stratigr 82:651–705

Campbell HJ (1994) The Triassic bivalves *Daonella* and *Halobia* in New Zealand, New Caledonia, and Svalbard. Inst Geol Nucl Sci Monogr 4:1–166

Caswell BA, Coe AL, Cohen AS (2009) New range data for marine invertebrate species across the early Toarcian (early Jurassic) mass extinction. J Geol Soc London 166:859–872

Chen J (1982) Mesozoic transgressions, regressions and bivalve provinces in China. Acta Geol Sin 21:334–346

Chinzei K (1982) Morphological and structural adaptations to soft substrates in the early Jurassic monomyarians *Lithiotis* and *Cochlearites*. Lethaia 15:179–197

Coates AG (1973) Cretaceous tethyan coral-rudist biogeography related to the evolution of the Atlantic Ocean. In: Hughes (ed) Organisms and continents through time. Spec Pap in Palaeontol 12:169–174

Cope JCW (2002) Diversification and biogeography of bivalves during the Ordovician period. In: Crame JA, Owen AW (eds) Paleobiogeography and biodiversity change: the Ordovician and Mesozoic-Cenozoic radiation. Geol Soc Spec Publ 194:25–52

Crame JA (1986) Late Mesozoic bipolar bivalve faunas. Geol Mag 123:611–618

Crame JA (1993) Bipolar molluscs and their evolutionary implications. J Biogeogr 20:145–161

Crame JA (1996) Evolution of high-latitude molluscan faunas. In: Taylor JD (ed) Origin and evolutionary radiation of the mollusca. Oxford University Press, Oxford

Crame JA (2002) Evolution of taxonomic diversity gradients in the marine realm: a comparison of late Jurassic and recent bivalve faunas. Paleobiology 28:184–207

Damborenea SE (1989) El género *Posidonotis* Losacco (Bivalvia, Jurásico inferior): su distribución estratigráfica y paleogeográfica. Actas 4° Congr Argent Paleontol y Bioestratigr (Mendoza, 1986) 4:45–51

Damborenea SE (1993) Early Jurassic South American pectinaceans and circum-Pacific paleobiogeography. Palaeogeogr Palaeoclimatol Palaeoecol 100:109–123

Damborenea SE (1996) Paleobiogeography of early Jurassic bivalves along the southeastern Pacific margin. 13° Congr Geol Argent y 3° Congr Explorac Hidrocarb (Buenos Aires). Actas 5:151–167

Damborenea SE (1998) The bipolar bivalve *Kolymonectes* in South America and the diversity of Propeamussiidae in Mesozoic times. In: Johnston PA, Haggart JW (eds) Bivalves: an eon of evolution—paleobiological studies honoring Norman D. Newell University Calgary Press, Calgary

Damborenea SE (2000) Hispanic Corridor: its evolution and the biogeography of bivalve molluscs. In: Hall RL, Smith PL (eds) Advances in Jurassic research 2000. Geo Res Forum 6:369–380

Damborenea SE (2002) Jurassic evolution of Southern Hemisphere marine palaeobiogeographic units based on benthonic bivalves. Geobios 35, MS 24:51–71

Damborenea SE, Manceñido MO (1979) On the palaeogeographical distribution of the pectinid genus *Weyla* (Bivalvia, Lower Jurassic). Palaeogeogr Palaeoclimatol Palaeoecol 27:85–102

Damborenea SE, Manceñido MO (1988) *Weyla*: semblanza de un bivalvo Jurásico andino. Actas 5° Congr Geol Chileno 2:C13–C25 (Santiago de Chile)

Darragh TA (1985) Molluscan biogeography and biostratigraphy of the Tertiary of southeastern Australia. Alcheringa 9:83–116

Dhondt AV (1992) Cretaceous inoceramid biogeography: a review. Palaeogeogr Palaeoclimatol Palaeoecol 92:217–232

Dhondt AV (1999) Palaeogeographical distribution patterns in Upper Cretaceous bivalves. Malacol Soc London, Biology and evolution of the bivalvia, Paper and Poster Abstracts: 18, Cambridge

Dickins JM (1993) Permian bivalve faunas. stratigraphical and geographical distribution. C R 12° Congr Internat Stratigr et Géol du Carbonifère et Permien 1:523–536

Douvillé H (1900) Sur la distribution géographique des Rudistes, des orbitolines at des orbitoides. Bull, Soc Géol France, 3° sér. 28:222–235

Duff KL (1978) Bivalvia of the English lower Oxford Clay (Middle Jurassic). Palaeontograph Soc Monogr 132(553):1–137

Emerson WK (1978) Mollusks with Indo-Pacific affinities in the eastern Pacific Ocean. Nautilus 92:91–96

Etter W (1996) Pseudoplanktonic and benthic invertebrates in the Middle Jurassic Opalinum Clay, northern Switzerland. Palaeogeogr Palaeoclimatol Palaeoecol 126:325–341

Flessa KW, Jablonski D (1995) Biogeography of recent marine bivalve molluscs and its implications for paleobiogeography and the geography of extinction: a progress report. Hist Biol 10:25–47

Fürsich FT, Sykes RM (1977) Palaeobiogeography of the European Boreal realm during Oxfordian (upper Jurassic) times: a quantitative approach. N Jb Geol Paläontol Adhand 172:271–329

Grant-Mackie JA, Aita Y, Balme BE, Campbell HJ, Challinor AB, MacFarlan DAB, Molnar RE, Stevens GR, Thulborn RA (2000) Jurassic palaeobiogeography of Australasia. In: Wright AJ, Young GC, Talent JA, Laurie JR (eds) Paleobiogeography of Australasian faunas and floras. Mem Assoc Australas Palaeontol 23:311–353

Hall CA (1964) Shallow water marine climates and molluscan provinces. Ecology 45:226–234

Hallam A (1967) The bearing of certain palaeozoogeographic data on continental drift. Palaeogeogr Palaeoclimatol Palaeoecol 3:201–241

Hallam A (1969) Faunal realms and facies in the Jurassic. Palaeontology 12:1–18

Hallam A (1971) Provinciality in Jurassic faunas in relation to facies and palaeogeography. In: Middlemiss FA, Rawson PF, Newall G (eds) Faunal provinces in space and time. Geol J Spec Issue 4:129–152

Hallam A (1977) Jurassic bivalve biogeography. Paleobiology 3:58–73

Hallam A (1981) Relative importance of plate movements, eustasy, and climate in controlling major biogeographical changes since the early Mesozoic. In: Nelson G, Rosen DE (eds) Vicariance biogeography: a critique. Columbia University Press, New York

Hallam A (1983) Early and mid-Jurassic molluscan biogeography and the establishment of the central Atlantic seaway. Palaeogeogr Palaeoclimatol Palaeoecol 43:181–193

Hallam A, Biró-Bagóczky L, Pérez E (1986) Facies analysis of the Lo Valdés Formation (Tithonian-Hauterivian) of the high cordillera of central Chile, and the palaeogeographic evolution of the Andean Basin. Geol Mag 123:425–435

Hayami I (1961) On the Jurassic pelecypod faunas in Japan. J Fac Sci, Univ Tokyo, Sect II. Geol Mineral Geogr Geophys 13:243–343

Hayami I (1969) Notes on Mesozoic "planktonic" bivalves. J Geol Soc Japan 75:375–385

Hayami I (1984) Jurassic marine bivalve faunas and biogeography in Southeast Asia. Geol Palaeontol Southeast Asia 25:229–237

Hayami I (1987) Geohistorical background of Wallace's Line and Jurassic marine biogeography. In: Taira A, Tashiro M (eds) Historical biogeography and plate tectonic evolution of Japan and Eastern Asia, Tokyo

Hayami I (1989) Outlook of the post-Paleozoic historical biogeography of pectinids in the Western Pacific region. Univ Mus Univ Tokyo Nat Cult 1:3–25

Hayami I (1990) Geographic distribution of Jurassic faunas in eastern Asia. In: Ichikawa K, Mizutani S, Hara I, Hada S, Yao A (eds) Pre-Cretaceous terranes of Japan. Publication of IGCP project 224, Osaka

Hillebrandt A (1980) Paleozoogeografía de Jurásico marino (Lías hasta Oxfordiano) en Suramérica. In: Zeil W (ed) Nuevos resultados de la investigación geocientífica alemana en Latinoamérica. Deuts Forschungs and Inst Colabor Cient, Tübingen

Hillebrandt A (1981) Kontinentalverschiebung und die paläozoogeographischen Beziehungen des südamerikanischen Lias. Geolog Runds 70:570–582

Jablonski D, Hunt G (2006) Larval ecology, geographic range, and species survivorship in Cretaceous mollusks: organismic versus species-level explanations. Am Nat 168:556–564

Jablonski D, Lutz RA (1983) Larval ecology of marine benthic invertebrates: paleobiological implications. Biol Rev 58:21–89

Jablonski D, Valentine JW (1990) From regional to total geographic ranges: testing the relationship in recent bivalves. Paleobiology 16:126–142

Jablonski D, Roy K, Valentine JW (1999) Dissecting the latitudinal gradient in marine bivalves. Malacol Soc London, Biology and evolution of the Bivalvia, Paper and Poster Abstracts: 24, Cambridge

Jablonski D, Roy K, Valentine JW (2000) Analysing the latitudinal gradient in marine bivalves. In: Harper EM, Taylor JD, Crame JA (eds) The evolutionary biology of the Bivalvia. Geol Soc Spec Publ 177:361–365

Jefferies R, Minton P (1965) The mode of life of two Jurassic species of "Posidonia" (Bivalvia). Palaeontology 8:156–185

Johannesson K (1988) The paradox of rockall: why is a brooding gastropod (Littorina saxatilis) more widespread than one having a planktonic larval dispersal stage (L. littorea)? Mar Biol 99:507–513

Kauffman EG (1973) Cretaceous Bivalvia. In: Hallam A (ed) Atlas of palaeobiogeography. Elsevier, Amsterdam

Kauffman EG (1975) Dispersal and biostratigraphic potential of Cretaceous benthonic Bivalvia in the Western Interior. In: Caldwell WGE (ed) The Cretaceous System in the Western Interior of North America. Spec Pap Geol Assoc Can 13:163–194

Kiessling W, Aberhan M (2007) Geographical distribution and extinction risk: lessons from Triassic-Jurassic marine benthic organisms. J Biogeogr 34:1473–1489

Knight RI, Morris NJ (1996) Inoceramid larval planktotrophy: evidence from the Gault Formation (Middle and basal Upper Albian), Folkestone, Kent. Palaeontology 39:1027–1036

Kobayashi T, Tamura M (1983a) On the oriental province of the Tethyan realm in the Triassic period. Proc Jpn Ac Ser B 59:203–206

Kobayashi T, Tamura M (1983b) The Arcto-Pacific Realm and the trigoniidae in the Triassic period. Proc Jpn Acad Ser B 59:207–210

Körner K (1937) Marine (Cassianer-Raibler) Trias am Nevado de Acrotambo (Nord-Peru). Palaeontogr A 86:145–237

Kříž J (1996) *Maida* nov. gen., the oldest known nektoplanktic bivalve from the Přídolí (Silurian) of Europe. Geobios 29:529–535

Krobicki M, Golonka J (2009) Palaeobiogeography of early Jurassic *Lithiotis*-type bivalve buildups as recovery effect after Triassic/Jurassic mass extinction and their connection with Asian palaeogeography. Acta Geoscient Sinica 30 supl 1:30–33

Liu C (1995) Jurassic bivalve palaeobiogeography of the proto-Atlantic and the application of multivariate analysis methods in palaeobiogeography. Beringeria 16:3–123

Liu C, Heinze M, Fürsich FT (1998) Bivalve provinces in the proto-Atlantic and along the southern margin of the Tethys in the Jurassic. Palaeogeogr Palaeoclimatol Palaeoecol 137:127–151

Liu C, Xie Y, Chen L (2007) Distribution of larval developmental types of marine bivalves along the eastern Pacific coast. Beringeria 37:95–103

Malchus N (2004) Early ontogeny of Jurassic bakevelliids and their bearing on bivalve evolution. Acta Palaeontol Pol 49(1):85–110

Marwick J (1953) Faunal migrations in New Zealand seas during the Triassic and Jurassic. N Z J Sci Technol B 34:317–321

Masse JP (1992) The Lower Cretaceous Mesogean benthic ecosystems: palaeoecologic aspects and palaeobiogeographic implications. Palaeogeogr Palaeoclimatol Palaeoecol 91:331–345

McRoberts CA (1997) Late Triassic North American halobiid bivalves; diversity trends and circum-Pacific correlations. In: Dickins JM et al (eds) Late Paleozoic and Early Mesozoic circum-Pacific events. Cambridge University Press, Cambridge, p 22

McRoberts CA, Aberhan M (1997) Marine diversity and sea-level changes: numerical tests for association using early Jurassic bivalves. Geol Runds 86:160–167

Nauss AL, Smith PL (1988) *Lithiotis* (Bivalvia) bioherms in the lower Jurassic of East-central Oregon, USA. Palaeogeogr Palaeoclimatol Palaeoecol 65:253–268

Newton CR (1983) Paleozoogeographic affinities of Norian bivalves from the Wrangellian, Peninsular, and Alexander terranes. In: Stevens CH (ed) Pre-Jurassic Rocks in Western North American suspect terranes. Pacific Section, Society of Economic Paleontologists and Mineralogists, Los Angeles

Newton CR (1987) Biogeographic complexity in Triassic bivalves of the Wallowa terrane, northwestern United States: Oceanic islands, not continents, provide the best analogues. Geology 15:1126–1129

Newton CR (1988) Significance of "Tethyan" fossils in the American Cordillera. Science 242:385–391

Niu Y, Jiang B, Huang H (2011) Triassic marine biogeography constrains the palaeogeographic reconstruction of Tibet and adjacent areas. Palaeogeogr Palaeoclimatol Palaeoecol 306:160–175

O'Foighil D (1989) Planktotrophic larval development is associated with a restricted geographic range in *Lasaea*, a genus of brooding, hermaphroditic bivalves. Mar Biol 103:349–358

Oschmann W (1993) Environmental fluctuations and the adaptive response of marine benthic organisms. J Geol Soc 150:187–191

Palmer CP (1989) Larval shells of four Jurassic bivalve molluscs. Bull Brit Mus Nat Hist Geol 45:57–69

Raby D, Laagdeuc Y, Dodson JJ, Mingelbier M (1994) Relationship between feeding and vertical distribution of bivalve larvae in stratified and mixed waters. Mar Ecol Progr Ser 103:275–284

Rey J, Andreo B, García-Hernández M, Martín-Algarra A, Vera JA (1990) The Liassic "Lithiotis" facies north of Vélez Rubio (Subbetic Zone). Rev Soc Geol España 3:199–212

Roy K, Jablonski D, Martien KK (2000) Invariant size-frequency distributions along a latitudinal gradient in marine bivalves. Proc Nation Acad Sci USA 97:13150–13155

Runnegar B (1975) Late Palaeozoic Bivalvia from South America: provincial affinities and age. An Acad Brasil Sci [1972] 44 (suppl):295–312

Runnegar B, Newell ND (1971) Caspian-like relict molluscan fauna in the South American Permian. Bull Am Mus Nat Hist 146:1–66

Sánchez MT, Babin C (2001) Paleogeographic distribution of Ordovician molluscan bivalves. In: International conference Paleobiogeogr Paleoecol, p 117

Savazzi E (1996) Preserved ligament in the Jurassic *Lithiotis*: apaptive and evolutionary significance. Palaeogeogr Palaeoclimatol Palaeoecol 120:281–289

Schatz W (2005) Palaeoecology of the Triassic black shale bivalve *Daonella*–new insights into an old controversy. Palaeogeogr Palaeoclimatol Palaeoecol 216:189–201

Scheltema RS (1977) Dispersal of marine invertebrate organisms: paleobiogeographic and biostratigraphic implications. In: Kauffman EG, Hazel JE (eds) Concepts and methods in biostratigraphy. Hutchinson and Ross, Stroudsburg, PA

Scheltema RS (1988) Initial evidence for the transport of teleplanic larvae of benthic invertebrates across the East Pacific barrier. Biol Bull 174:145–152

Scheltema RS, Williams IP (1983) Long-distance dispersal of planktonic larvae and the biogeography and evolution of some Polynesian and Western Pacific mollusks. Bull Mar Sci 33:545–565

Scotese CR (1997) Paleogeographic Atlas. PALEOMAP Progress Report 90-0497, Department of Geology, University of Texas at Arlington, Arlington, Texas

Sha J (1996) Antitropicality of the Mesozoic Bivalves. In: Pang ZH et al (eds) Advances in Solid Earth Sciences. Science Press, Peking

Sha J (2002) Hispanic corridor formed as early as Hettangian: on the basis of bivalve fossils. Chin Sci Bull 47:414–417

Shi GR, Grunt TA (2000) Permian Gondwana-Boreal antitropicality with special reference to brachiopod faunas. Palaeogeogr Palaeoclimatol Palaeoecol 155:239–263

Shurygin BN (2005) Biogeografiya, fatsii i stratigrafiya nizhnej i srednej Yury Sibiri po dvustvorchatym mollyuskan [Lower and Middle Jurassic biogeography, facies and stratigraphy in Siberia based on bivalve mollusks]. Trofimuk United Institute of Geology, Geophysics and Mineralogy; Institute of Petroleum Geology. Academic Publishing House "Geo", pp 156 Novosibirsk

Silberling NJ (1985) Biogeographic significance of the Upper Triassic bivalve *Monotis* in Circum-Pacific Accreted Terranes. In: Howell DG (ed) Tectonostratigraphic Terranes of the Circum-Pacific region. Circum-Pacific Council for Energy and Mineral Resources, Earth Science Series 1:63–70

Silberling NJ, Grant-Mackie JA, Nichols KM (1997) The Late Triassic Bivalve *Monotis* in Accreted Terranes of Alaska. US Geol Surv Bull 2151:1–21

Skelton PW, Wright VP (1987) A Caribbean rudist bivalve in Oman: island-hopping across the Pacific in the Late Cretaceous. Palaeontology 30:505–529

Skwarko SK (1983) *Somareoides hastatus* (Skwarko), a new Late Triassic bivalve from Papua New Guinea. Bull Bur Min Res Geol Geophy-s 217:67–68

Smith AG, Briden JC (1977) Mesozoic and Cenozoic paleocontinental maps. Cambridge University Press, Cambridge

Smith PL, Westermann GEG, Stanley GD Jr, Yancey TE Jr (1990) Paleobiogeography of the Ancient Pacific (response by Newton CR). Science 249:680–683

Speden IG, Keyes IW (1981) Illustrations of New Zealand Fossils. New Zealand Department of Scientific and Industrial Research, DSIR Information Series, Wellington, 150

Stanley SM (1972) Functional morphology and evolution of byssally attached bivalve mollusks. J Paleontol 46(2):165–212

Stevens GR (1967) Upper Jurassic fossils from Ellsworth Land, West Antarctica, and notes on Upper Jurassic biogeography of the South Pacific region. N Z J Geol Geophys 10:345–393

Stevens GR (1977) Mesozoic biogeography of the South-West Pacific and its relationship to plate tectonics. In: International Symposium on the Geodynamics of the SW Pacific. Ed. Technip, Paris

Stevens GR (1980) Southwest Pacific faunal palaeobiogeography in Mesozoic and Cenozoic times: a review. Palaeogeogr Palaeoclimatol Palaeoecol 31:153–196

Tamura M (1990) The distribution of Japanese Triassic bivalve funas with special reference to parallel distribution of inner arcto-Pacific fauna and outer Tethyan fauna in Upper Triassic.

In: Ichikawa K, Mizutani S, Hara I, Hara S, Yao A (eds) Pre-Cretaceous terranes of Japan. Publ IGCP Project 224:347–359

Tanoue K (2003) Larval ecology of Cretaceous inoceramid bivalves from northwestern Hokkaido, Japan. Paleontol Res 7:105–110

Tausch von Gloeckelsthurn L (1890) Zur Kenntniss der Fauna der "grauen Kalke" der Süd-Alpen. Abhandl k.k. Geolog Reichs 15(2):1–42

Voigt S, Hay WW, Höfling R, De Conte RM (1999) Biogeographic distribution of late Early to Late Cretaceous rudist-reefs in the Mediterranean as climate indicators. Geol Soc Am Spec Pap 332:91–103

Wignall PB (1990) Observations on the evolution and classification of dysaerobic communities. In: Miller W (ed) Paleocommunity temporal dynamics: the long-term development of multispecies assemblies. Paleontol Soc Spec Publ 5:99–111

Wignall PB, Simms MJ (1990) Pseudoplankton. Palaeontol 33:359–378

Zinsmeister WJ (1979) Biogeographic Significance of the late Mesozoic and early Tertiary Molluscan Faunas of Seymour Island (Antarctic Peninsula) to the final breakup of Gondwanaland. In: Gray J, Boucot AJ (eds) Historical biogeography, plate tectonics, and the changing environment. Oregon State University Press, Oregon

Zinsmeister WJ (1982) Late Cretaceous-early Tertiary molluscan biogeography of the southern circum-Pacific. J Paleontol 56:84–102

Chapter 4
Regional Scale

Abstract First-hand knowledge of Triassic and Jurassic bivalve faunas from western South America can be applied to the discussion of several regional issues related to paleobiogeography, such as the close relationship of bivalve local distribution with sedimentary facies. Special attention is paid to the analysis of bivalve species distribution along a 25° latitudinal range of the Panthalassa paleocoast, done in time slices corresponding to the Jurassic stages. The results did not show the expected decrease in species diversity toward higher latitudes. A local diversity increase between 30 and 40° S Lat., especially evident for Pliensbachian and Toarcian times, may be due to the establishment of a variety of habitats within the extensive Neuquén basin. The latitudinal distribution pattern was also explored using cluster analysis and faunal turnover analysis. These two analytical methods, together with the distribution of species with high-latitude biogeographic affinities, consistently show the existence of a faunal turnover area, here interpreted as the boundary between Tethyan and South Pacific faunas, which migrated southwards with time. This shift amounts to about 700 km (about 8° paleolatitude) between Hettangian and Toarcian times. The analysis for changes in proportion of different superfamilies along this latitudinal range indicated a decreasing diversity trend toward higher latitudes for Trigonioidea, Limoidea, and Pholadomyoidea, while the opposite trend was shown for some of the time slices by Nuculanoidea, Monotoidea, Pectinoidea, and Crassatelloidea.

Even at a regional scale, paleobiogeographic studies are complex and it is necessary to have good paleogeographic base maps and a reliable and independently based time frame. Unfortunately, this is not the case for several regions included in this study, in many areas of the Southern Hemisphere the geologic history is not known in sufficient detail yet. For this reason, and also because it is the area we know best, we will mostly exemplify this approach using data from western Argentina/Chile (Fig. 4.1).

S. E. Damborenea et al., *Southern Hemisphere Palaeobiogeography of Triassic-Jurassic Marine Bivalves*, SpringerBriefs Seaways and Landbridges: Southern Hemisphere Biogeographic Connections Through Time, DOI: 10.1007/978-94-007-5098-2_4, © The Author(s) 2013

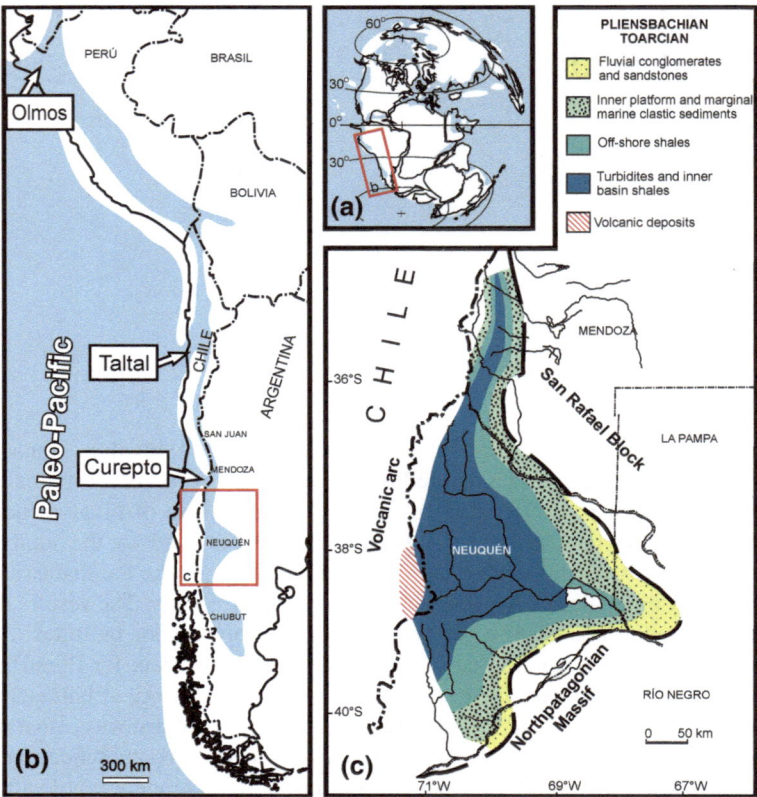

Fig. 4.1 General paleogeographic context for the late Early Jurassic at different scales. **a** Global scale, compiled from several sources. **b** Continental scale, showing western South American paleogeography (modified from Vicente 2005). **c** Regional scale, generalized paleogeography of the Neuquén Basin during Pliensbachian-Toarcian times (modified from Legarreta and Uliana 2000). The extent and distribution of the different paleoenvironments changed with time, see discussion in text. Note that in **a** the inferred paleolatitudes are indicated, while in **c** present-day latitudes are referred to instead

4.1 Facies and Bivalve Distributions: Examples from the Neuquén Basin

The Mesozoic South American basin system developed along the eastern Panthalassa margin. By Norian times these basins probably had no less than two connections with the open ocean, while by Early Jurassic times at least three passages were established (Vicente 2005; See Fig. 4.1b), the southernmost (Curepto) connecting the Neuquén basin with the Paleo-Pacific. The Neuquén Basin extended at the western margin of South America; it formed a wide eastward embayment between 36 and 40° present S latitude. It was a typical back-arc basin, and its paleogeography has been the subject of many studies since the pioneer synthesis by Groeber (1946). Paleogeographic maps depicting the extension and facies distribution of marine deposits in the basin are now available at stage-scale intervals (Gulisano 1992; Riccardi et al. 1992, 2011; Legarreta and Uliana 1996, 2000).

This local basin mostly developed in the present territories of San Juan, Mendoza, and Neuquén provinces in Argentina and in part of central Chile (Fig. 4.1c), and accumulated a thick marine sedimentary succession from late Triassic to late early Cretaceous times. During the late Early Jurassic it extended temporarily to the South into Chubut province in Argentina.

The history of this extensive basin was complex but is relatively well known (see Legarreta and Uliana 1996 for a good synthesis), in part due to the economic interests derived from oil production. As a result, the biostratigraphic (Riccardi 2008a, b), paleogeographic, and paleoenvironmental frames are good enough to attempt the analysis and interpretation of regional bivalve distribution.

The available paleogeographic maps allow us to examine the distribution of bivalves in the context of the environmental conditions at each stage. The gathered information on the detailed geographic range of more than a hundred Triassic and Jurassic bivalve species from the Neuquén Basin demonstrates once again the well-known fact that lithofacies (mainly related to substrate type) and distribution are highly correlated, although this is not the only factor to account for. Other aspects, such as water circulation patterns, oxygenation, paleosalinity, and paleotemperature, should also be considered. Of the wealth of data at hand, only a few examples, placed at different time intervals, will be mentioned to exemplify the importance of an adequate evaluation of facies when dealing with paleogeographic distributions of bivalves.

4.1.1   Late Triassic

An interesting example at a large regional scale derives from the affinities of the late Triassic fauna from southern Mendoza in Argentina (Damborenea and Manceñido 2012). Many of its bivalve species (e.g., belonging to the genera *Cassianella, Palaeocardita, Septocardia, Minetrigonia?, Liostrea*) are more related to those known from the late Triassic of Perú and northern Chile than to those from the Chilean coast at the same latitude (Curepto area). There is an evident environmental distinction between the two sets of species, a western belt being dominated by open ocean species (probably outer shelf to off-shore), mostly monotoids and limids. On the other hand, the eastern fauna developed in a marine, well-oxygenated, littoral environment, with shallow water depth and type of substrate as the critical factors limiting faunal distribution. It is evident here that both main facies are widely distributed latitudinally in western South America as two belts roughly parallel to the paleoshore (Fig. 4.2).

Fig. 4.2 Main facies and faunal distribution for latest Triassic (Norian–Rhaetian) bivalve bearing deposits in western South America. Only main areas are mentioned, see discussion in text

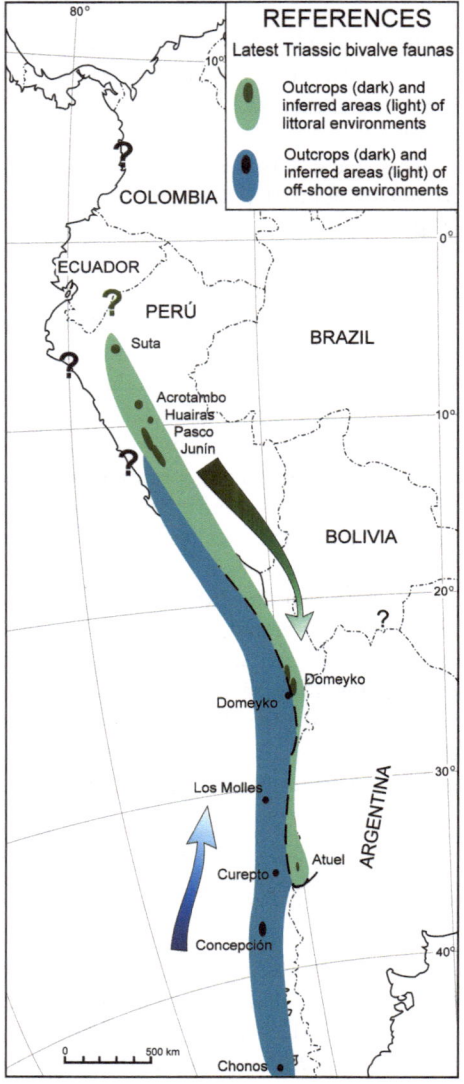

Goodwin (1997) discussed a similar situation in the Northern Hemisphere: Norian faunas from Sonora (Mexico) are very close to those from Nevada (USA) despite latitudinal differences in present-day location. Only after an adequate evaluation of paleoenvironmental conditions a discussion of paleogeographic affinities can be attempted. In our case, littoral faunas are related to northern ones, while open marine faunas have high-latitude affinities, in agreement with inferred surface paleocurrents (see Sect. 1.6, Fig. 1.3a).

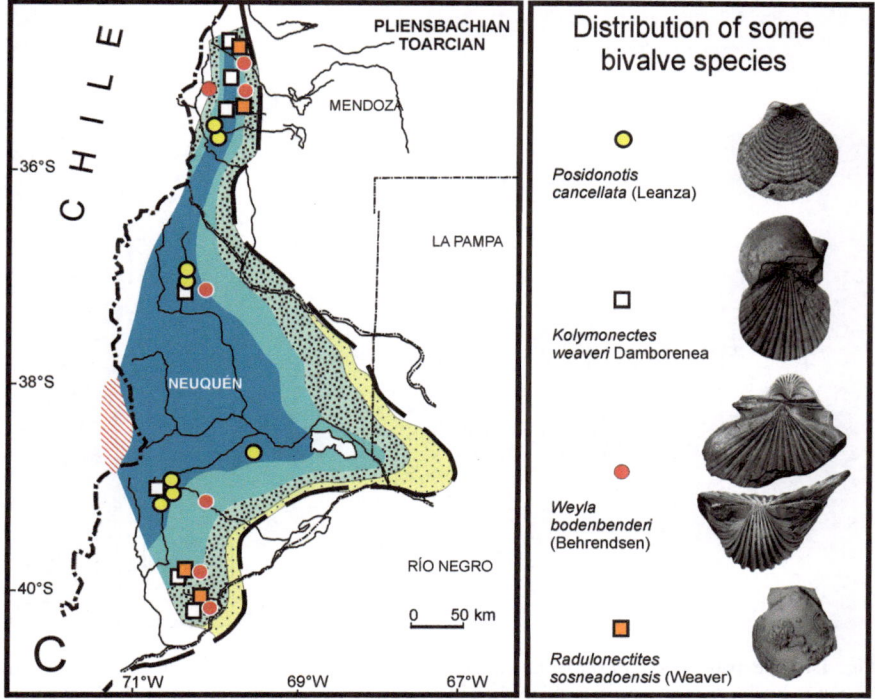

Fig. 4.3 An example of the close relationship between faunal distribution and paleogeography in the Neuquén Basin. The localities where four Pliensbachian-Toarcian pectinoid species occur are indicated on the corresponding paleogeographic map (paleogeography modified from Legarreta and Uliana 2000). Specimens figured not to scale. References to facies in Fig. 4.1

4.1.2 Early Jurassic

The analysis of the distribution of four pectinoid species during the Pliensbachian-Toarcian within the Neuquén embayment (Fig. 4.3) again shows that there is a close correspondence between distribution and environment. The thin-shelled and probably swimming species *Posidonotis cancellata* (Leanza) and *Kolymonectes weaveri* Damborenea are only abundant in inner basin areas, although they may occasionally appear elsewhere. On the other hand, the recliner *Weyla bodenbenderi* (Behrendsen) and the byssate *Radulonectites sosneadoensis* (Weaver) are usually found in sublittoral to platform sites.

In this context, and extending beyond the region, it is interesting to further discuss the distribution of species of the peculiar genus *Posidonotis*. This pectinoid clearly qualifies as a "paper-clam" or "flat-clam" and occurs abundantly in dark shales (Fig. 4.4), in facial equivalents to the "Posidonienschiefer" of central Europe. In South America, the genus ranged from latest Pliensbachian to earliest Toarcian times (Damborenea 1989), forming highly concentrated monospecific

Fig. 4.4 The paper-clam *Posidonotis cancellata* (Leanza), field photograph at Arroyo Lapa, Neuquén Basin, earliest Toarcian. Slab showing the usual preservation as dense shell pavements in dark shales. *Inset* nearly complete valve

shell pavements in thin-bedded sediments related to the Early Toarcian anoxic event (Al-Suwaidi et al. 2010). The species *P. dainellii* Losacco is known from Japan in beds of the same age (Hayami 1988). In North America, *Posidonotis* is known from older deposits (Sinemurian), concentrated at a few levels also probably related to poorly oxygenated bottom conditions (Aberhan and Pálfy 1996). *Posidonotis* was apparently absent from extreme Boreal and Austral regions, and was restricted to low and mid-paleolatitudes of the circum-Pacific, sporadically occurring in the Tethys (Damborenea 1989; Aberhan and Pálfy 1996). Throughout its geographic range, *Posidonotis* thrived in dysaerobic conditions (see Sect. 3.2.5, Figs. 3.1 and 3.2), but it could survive too in more oxygenated environments, where it was never abundant.

The distribution of *Kolymonectes weaveri* Damborenea also extended beyond the Neuquén embayment, since this propeamussiid with clear bipolar affinities was very abundant in several localities further South, in Chubut Province as well (Fig. 4.5). The genus did not extend to low latitudes although the right facies do occur further north, but it is also known from late Triassic-Early Jurassic deposits in the Northern Hemisphere mid- and high-latitude regions (Damborenea 1998, see references therein). In Argentina, the species occurs in very fine-grained, mostly light-colored fine-grained sandstones and shales, and is rare in dark shales; it

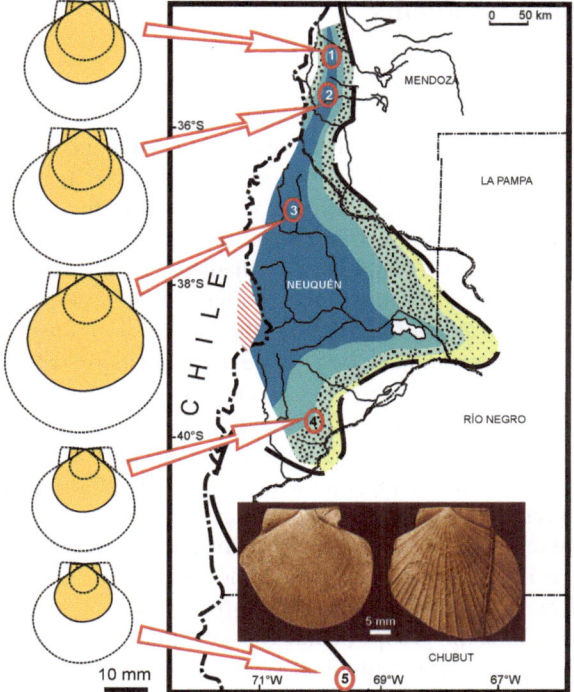

Fig. 4.5 Mean size attained by left valves of *Kolymonectes weaveri* Damborenea at different regions (1–5) of the Neuquén Basin: *1* Atuel/Portezuelo Ancho, *2* Puchenque/Serrucho, *3* Chacay Melehue, *4* La Pintada, *5* Agnia. Only complete specimens were measured. Mean size: shaded, minimum and maximum size shown as *broken lines*. *Inset* left and right valves of complete specimen (MLP 23807), Pliensbachian of Chacay Melehue, Neuquén. Right valve shows shallow byssal notch and the ventral margin broken along the boundary of inner and outer shell layers. References to facies in Fig. 4.1 (Modified from Damborenea 1998)

clearly preferred well-oxygenated, open sea conditions. The abundance of the local species increases to the south (southernmost Neuquén and Chubut). Populations from different localities differ greatly in mean shell size but not latitudinally. It is probably related to depth and/or oxygen availability rather than temperature or other latitude-related factors. Larger sizes are attained in deposits of low water energy, the only accompanying fauna being ammonites. Young individuals were probably byssally attached but the byssal notch became obsolete in adults and was then probably not functional; adults most likely lived resting on the substrate and might have been good occasional swimmers. It is also interesting that propea-mussiids (or "glass scallops") in general seem to be relegated nowadays to safe places in the deep sea (Waller 2011), whereas they clearly inhabited shallower waters in the Triassic and Jurassic. They are also regarded as living relics due to their primitive soft anatomy and shell microstructure (Waller 1971, 2006).

4.2 Latitudinal Gradients

Bivalves have been fundamental to investigate the nature and origin of marine diversity gradients (Crame 2000a, b; Valentine and Jablonski 2010, and references therein). One of the most obvious relationships between biogeography and ecology is the existence of latitudinal gradients in species diversity. This is generally understood to mean an increase in species richness from the poles to the tropics, and is well documented for terrestrial faunas. This pattern was extrapolated to the marine biotas (Sanders 1968), and later verified for prosobranch gastropods (Roy et al. 1998) and bivalves, both from the continental shelf (Jablonski et al. 2000) and deep sea (Rex et al. 1993, 2000) of the Northern Hemisphere. Latitudinal gradients are thus a remarkable large-scale biotic pattern, which is shared by terrestrial and marine organisms. Roy et al. (1998) analyzed various previous hypothesis to explain the origin of this major pattern, and, based on marine gastropods living on the shelves of the eastern Pacific and western Atlantic, they concluded that sea surface temperature (as the result of solar energy input) is significantly correlated to the strikingly similar latitudinal gradients observed.

Marine bivalves show clear latitudinal diversity gradients as well, but the observed pattern is not simple and it does not appear to be symmetric in both hemispheres (Rex et al. 1993; Crame 2000a, b). Even an inverse gradient is also regionally known for some groups (Valdovinos et al. 2003; Kindlmann et al. 2007) along the Chilean coast. The latitudinal gradients in biodiversity are not easy to interpret because they are strongly influenced by local conditions and the history of the regions concerned (Crame 2000b), but it was proposed that it is maintained by high tropical origination rates (Valentine and Jablonski 2010).

The various hypotheses proposed to explain the origin of this pattern are of general nature and thus imply that this feature should have been present in past biotas as well. Crame (2000a, b) proved that latitudinal gradients in bivalve taxonomic diversity can be traced back to the late Paleozoic in both hemispheres, though they were not symmetric. Furthermore, he observed that late Paleozoic and late Jurassic diversity gradients were weaker than present ones (Crame 2001, 2002), and there was a dramatic increase in these gradients during the Cenozoic.

Apart from this well-known and universally recognized latitudinal gradient in diversity, knowledge about other latitudinal gradients (related for instance to taxonomy, functional groups, size or intraspecific variability) is still patchy, but again bivalves provide good data for their discussion.

Based on the analysis of latitudinal distribution of eastern Pacific bivalves, Roy et al. (2000) concluded that latitudinal patterns of species richness are decoupled from patterns of *body size* (modal size, average size and size range). Size-frequency distributions are significantly undistinguishable over more than 5,000 km of continental shelf, from the equator to the polar sea, although the family level composition of the modal size class varies considerably with latitude.

On the other hand, concerning bivalve *taxonomy*, Crame (2000a) argued that the steepest latitudinal biodiversity gradients for bivalves are related to the youngest clades. Thus, the present-day latitudinal gradient in marine bivalves is influenced by a tropical and low latitude concentration of infaunal taxa (mainly heteroconchs), while the gradient in epifaunal pteriomorphs is far less marked. These results were compared with a similar analysis of late Jurassic bivalve distribution (Crame 2002), and the differences observed were attributed to a large Cenozoic heteroconch diversification, which caused a steepening of the latitudinal gradient, more evident in the Northern Hemisphere. The Anomalodesmata are peculiar in this context, since they lack a diversity maximum in the tropics (Krug et al. 2007), and have two maxima in temperate Northern and Southern Hemispheres.

The latitudinal diversity gradients in living marine bivalves are also evident for both infauna and epifauna, and for most major *functional groups* (Roy et al. 2000), except for the deposit-feeding protobranchs.

Latitudinal diversity gradients in marine Mesozoic bivalves have been recognized and described at a global scale, but little is known about regional patterns in detail. The paleogeographic position of South America during the Early Jurassic and the availability of outcrops make this part of Gondwanaland a privileged place to study the latitudinal variation of benthonic faunas. The South American Paleo-Pacific margin had a wide range of paleolatitudes, and the available data allow the examination of bivalve biodiversity (species richness) along a strip of over 20° paleolatitude (presently between 20 and 45° S lat.).

The first results of such an analysis (Damborenea 1996) showed that within that paleolatitudinal range there was no apparent diversity gradient referable to latitude for three different Early Jurassic time intervals then considered. On the other hand, the latitudinal distribution of the species strongly suggested the presence of a wide area of mixed faunas between Tethyan (to the North) and Austral (to the South) faunas, a subject which will be more extensively discussed later (Sect. 5.4).

A new analysis performed on the basis of an updated database (sources listed in Sect. 5.1.1) of the bivalve species distribution provided the following results, discussed here in general and then discriminated by systematic and paleobiogeographic affinities (in the sense discussed in Sect. 3.3).

4.2.1 Diversity gradients

The distribution of bivalve species in about 200 localities from Chile and Argentina between 20 and 45° (Table 4.1) was recorded for the four Early Jurassic stages: (a) Hettangian (Fig. 4.7), (b) Sinemurian (Fig. 4.9), (c) Pliensbachian (Fig. 4.11), and (d) Toarcian (Fig. 4.13). Though it is possible to analyze shorter time intervals for the distribution of Argentinean Early Jurassic bivalves from the Neuquén Basin, where their time ranges are determined accurately by accompanying ammonites, the same precision is not yet possible for some of the other areas. The Neuquén Basin time ranges cannot be extrapolated to the whole area since differences may be expected due to the large geographic distances involved.

Table 4.1 Main localities and data sources for the latitudinal analysis along western South America. Data are arranged according to 13 areas with a 2° present-day latitudinal range each

Region	Present-day latitude range	Main localities	Sources of data
0	20–22°	Socosani, Longacho, Pampa Soledad, Quillagua	Pérez and Reyes 1994; Pérez et al. 2008
1	22–24°	Cerritos Bayos, Moctezuma, Limón Verde, Calama, Caracoles, Cuevitas, Cochrane, Azabache	Steinmann 1881; Möricke 1894; Pérez and Levi 1961; Harrington 1961; Pérez and Reyes 1977, 1994; Hillebrandt 1990; Aberhan 1994; Pérez et al. 2008
2	24–26°	Domeyko, Alto Varas, Bonita, Chaco Sur, Incahuasi, Tres Hidalgos, Oreganito, Burros, Aspera, Profeta, Mulas, Paposo, Vaquillas Altas, Carretas, Cachina	Hillebrandt 1971, 1973, 1977, 1980, 2000; Pérez and Reyes 1977, 1994; Covacevich and Escobar 1979; Chong and Hillebrandt 1985; Hillebrandt et al. 1986; Quinzio 1987; Aberhan 1994; Pérez et al. 2008
3	26–28°	Doña Inés Chica, Minillas, Pan de Azúcar, Pedernales, San Juan, Asientos, Caballo Muerto, Tamberías, El Peñón, La Chaucha, Paipote, El Bolito, El Patón, El Carbón, Cortaderita, La Ternera, Potrerillos, Vaca Muerta, Yerbas Buenas, San Pedrito, Larga, Noria, Llareta, San Miguel, Figueroa, Jorquera, La Guardia, Calquis, Los Eucaliptus, Las Vizcachas, Las Trancas	Möricke 1894; Philippi 1899; Hillebrandt 1973; Pérez and Reyes 1977; Hillebrandt and Schmidt-Effing 1981; Mercado 1982; Sepúlveda and Naranjo 1982; Chong and Hillebrandt 1985; Hillebrandt and Westermann 1985; Hillebrandt et al. 1986; Quinzio 1987; Hillebrandt 1990, 2000; Aberhan 1992, 1993, 1994, 2004; Pérez et al. 1995, 2008; Aberhan and Hillebrandt 1996
4	28–30°	Manflas, Amolanas, La Iglesia, Pulido, Juntas del Tolar, Salto del Toro, El Tránsito, El Corral, La Totora, Chanchoquín, Paitepén, Plaza, Tatul, Las Pircas, Pinte, Picudo, La Plata, La Papa, Los Cuartitos, Calabocito, La Punilla, Doña Ana, Elqui	Bayle and Coquand 1851; Burmeister and Giebel 1861; Möricke 1894; Philippi 1899; Groeber et al. 1953; Thiele 1964; Hillebrandt 1971, 1973, 1977, 2002; Pérez and Reyes 1977; Hillebrandt and Westermann 1985; Aberhan 1992, 1994, 2004; Pérez et al. 1995, 2008; Aberhan and Hillebrandt 1996, 1999
5	30–32°	Matahuaico, Tres Cruces, Mostazal, Los Pingos, Los Erizos, El Pachón	Bayle and Coquand 1851; Conrad 1855; Philippi 1899; Dediós 1967; Mpodozis et al. 1973; Pérez and Reyes 1977; Ramos et al. 1993; Aberhan 1994, 2004; own data

(continued)

Table 4.1 (continued)

Region	Present-day latitude range	Main localities	Sources of data
6	32–34°	Las Flores, La Honda, Los Molles, La Laguna, La Ligua, del Pobre	Rigal 1930; Thomas 1958; Cecioni and Westermann 1968; Pérez and Reyes 1977; Volkheimer et al. 1978; Damborenea 1987a, b, 2002a; Pérez et al. 2008; own data
7	34–36°	La Manga, Malo, La Horqueta, Tinguiririca, Blanco, Pedrero, Los Caballos, Las Chilcas, Araya, La Brea, La Bajada, Curepto, Portezuelo Ancho, Deshecho, Santa Elena, Salado, Troncoso, El Infiernillo, Serrucho, Puchenque, Tricolor, Barda Blanca, Chacayco, Poti-Malal	Behrendsen 1891; Philippi 1899; Jaworski 1925; Groeber et al. 1953; Damborenea 1987a, b, 2002a, 2004; Riccardi et al. 1988, 1991; Pérez et al. 1995; Damborenea and Lanés 2007; own data
8	36–38°	Los Baños, Tocuyo, Ñiraico, Rajapalo, Perfil, Lista Blanca, Chacay Melehue,	Damborenea 1987a, b, 2002a; own data
9	38–40°	Del Gringo, Los Toldos, Ñireco, Pichi Picún Leufú, Granito, Puruvé Pehuén, Vuta Picún Leufú, Lonqueo, Ibáñez, Espinazo del Zorro, Llao–Llao, Aluminé, Lapa, Charahuilla, Keli Mahuida, Los Molles, Picún Leufú, La Jardinera, Catán Lil, Santa Isabel	Weaver 1931; Groeber et al. 1953; Damborenea 1987a, b, 2002a; Pérez et al. 1995; own data
10	40–42°	Carrán Cura, Salitral Grande, Sañicó, Los Chilenos, Los Pantanos, Roth, Mesa, La Pintada, del Vasco, Corona, Piltriquitrón	Leanza 1942; Manceñido and Damborenea 1984; Damborenea 1987a, b, 2002a; Pérez et al. 1995; own data
11	42–44°	Gualjaina, Pescado, Cuche, Peña, La Carlota, Agnia, Currumil, Nahuelquir, Chapingo, Carnerero, Plate, Lomas Chatas, El Córdoba	Piatnitzky 1936; Robbiano 1971; Lesta et al. 1980; Lage 1982; Nullo 1983; Benito and Chernicoff 1986; Vizán 1988; Massaferro 2001; own data
12	44–46°	Puelman, Negro, Altamirano, Piedra Shotle, Parra, Betancourt, La Trampa, Nueva Lubecka, Aguada Loca, Ferrarotti, Loncopán, Salazar, Guadal, Colorado	Piatnitzky 1933, 1936; Feruglio 1934; Wahnish 1942; Robbiano 1971; Malumián and Ploszkiewicz 1976; Blasco et al. 1980; Lesta et al. 1980; Nullo 1983; Cortiñas 1984; Pérez et al. 1995; Damborenea 2002a; Pagani et al. 2012; own data

Table 4.2 Number of bivalve species present in each latitudinal segment for each of the four stages considered

Regions	0	1	2	3	4	5	6	7	8	9	10	11	12
Hettangian	1	6	7	12	0	0	1	18	–	–	–	–	–
Sinemurian	1	8	10	23	54	3	3	67	1	–	–	–	–
Pliensbachian	–	5	2	84	57	6	24	87	32	50	67	14	50
Toarcian	–	3	1	30	55	3	34	54	24	23	3	4	8

Regions numbered as in Table 4.1

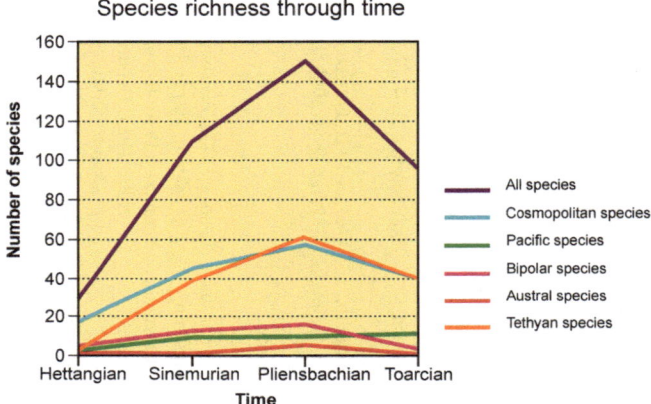

Fig. 4.6 Total bivalve species richness through time for the whole latitudinal range considered in western South America (20–46° present-day S latitude)

Only eight localities belong to the Coastal Cordillera of Chile; all the others are in the Andes.

The database for the analysis is a species list showing the distribution of 233 bivalves in 13 areas (0–12), each with a latitudinal range of 2°, spanning a north–south strip from 20° to 46° S latitude. This database also includes information on each species taxonomic and paleobiogeographic affinities. A summary of data is shown on Table 4.2. Though the purpose is the consideration of paleobiogeographic issues, data were initially plotted on their present-day positions to avoid a priori bias and circular reasoning. As pointed out by Rosen (1992), present-day positions are the only universally objective reference for fossil locations available so far.

It was preferred at this stage to use all the available information, regardless of the paleoecologic types of the bivalve species. These comprise several life-habit types, though most of them are shallow platform dwellers. It is well known that facies control may significantly affect distribution of some bivalves (see Sect. 4.1), and this should be distinguished from regional factors determining provincialism.

Nevertheless, it is thought that the large number of records and localities taken into account (comprising a wide range of facies within each area) make this "noise" factor less of a problem.

In central Chile and Argentina the occurrence of typically Tethyan bivalves in the same area as high-latitude bivalve species is a consequence of a mid-latitude paleoposition of this region during the Early Jurassic without significant barriers along the East Paleo-Pacific margin, the probable pantropic nature of Tethyan faunas, and a shallow sea connection with the western Tethys from middle Early Jurassic times onwards (Damborenea 1993).

Some general trends through the time involved will be discussed first, before treating the paleolatitudinally related features of the faunas. Within the study area, there is a slight decrease in the percentage of "local" species through time from the Hettangian (67 %) to the Toarcian (60 %). This decline of restricted species during the Early Jurassic is in agreement with similar trends observed in several areas of the northern hemisphere (see Hallam 1977) for endemic bivalve genera. It is interesting to note that Hallam (1977, Fig. 2) recorded an opposite trend for South America but then correctly attributed it to poorly documented data.

We can now add plots of overall bivalve diversity (number of species) through time (Fig. 4.6) along the whole studied area in western South America, which shows a sharp maximum in the Pliensbachian. This fact is in agreement with plots of the number of bivalve genera worldwide along this same time interval (Hallam 1977, Fig. 1). When data are discriminated according to the biogeographic affinities of each species, it is evident that all types participate in this diversity increase except "Pacific" species, which are equally numerous in Sinemurian, Pliensbachian, and Toarcian times. For this reason, there is almost no change in the proportion with which these types contribute to the general composition of faunas through time.

For living bivalves taxonomic diversity at family, genus, and species levels are covariant with latitude (Stehli et al. 1967; Stehli 1968), and this can be extrapolated to fossil faunas, even during times when climatic belts were apparently ill-defined (Stehli et al. 1969) as seems to have been during the Early Jurassic.

Concerning diversity latitudinal gradients, the data do not show the expected decrease in species diversity toward higher latitudes in the geographic range considered here, but, instead, a local diversity increase between 30° and 40°, which is especially evident for Pliensbachian and Toarcian times. This local increase may be due to the establishment of favorable conditions and an increased variety of habitats within the extensive Neuquén Basin, which at that time was a quasi-isolated shallow water epeiric sea (see Sect. 4.1 for paleogeographic reconstructions of this area).

4.2.2 Latitudinal Distribution of Biogeographic Affinities Through Time

Latitudinal species ranges could be sensitive to the variety of physical conditions displayed through the stretch of coast, including water current systems, geomorphology of the coast, input of freshwater, local oxygenation conditions, and so on. At the scale of our data, discontinuities are recognizable by the concurrence of latitudinal breaks of different species. Our data set can be analyzed quantitatively to investigate first the presence of latitudinal locality groups characterized by their fauna. The northern and southern limits of species ranges were analyzed to recognize sudden changes in faunal composition (see description of methodology applied in Sect. 2.3.1). Data were then discriminated by superfamilies and also by paleobiogeographic affinities, to complete the characterization of the latitudinal distribution patterns for each of the time intervals. In this way, the evolution of the observed patterns through the Early Jurassic (c. 25 Ma) can be described as well.

4.2.2.1 Hettangian

No deposits of this age bearing marine bivalves are known to the south of 36° S, and thus the analysis is constrained to the northern regions of our range (Fig. 4.7).

Cluster analysis for the Hettangian (Fig. 4.8) shows certain latitudinal gradient, discriminating between northern (20°–28° S) and southern (32°–36° S) bivalve faunas. Although northern and southern limits of distribution show some turnover between 26° and 28° S, it must be pointed out that there is no data for the latitudes between 28° and 32° S, so the peak for southern limits of distribution may be overestimated.

With reference to the biogeographic affinities of the analyzed species (Fig. 4.15a), there are no strictly significant changes along this latitudinal gradient, but there is a relative reduction toward the south of species with Pacific affinities (linear predictor: -0.48) that is in the limit of significance ($p = 0.083$); this biogeographic grouping disappears by 28° S. Species with high-latitude (bipolar plus austral) affinities are present and relatively abundant through the whole of the range (Fig. 4.15a). They also show a not strictly significant ($p = 0.098$) trend, with their proportion increasing slightly and evenly toward the South (linear predictor: 0.12), whereas in absolute numbers the maxima are at the intervals 26–28° (3 species) and at 34–36° (7 species). Species of this group which reach the northernmost regions are: *Palmoxytoma* cf. *cygnipes* (Young and Bird) and *Agerchlamys?* sp. The southern diversity peak for high-latitude taxa also contain *Palmoxytoma* cf. *cygnipes*, but to these, *Otapiria pacifica* Covacevich and Escobar, and species of *Parainoceramus?*, *Kalentera* and Astartidae are added.

Fig. 4.7 Latitudinal ranges of Hettangian bivalve species (each vertical line represents one species), discriminated by paleobiogeographic affinities. The 13 latitudinal areas used for this analysis (spanning 2° each) are numbered, and each locality is represented by a *black dot*. Notice that localities south of 36° do not bear Hettangian faunas. Extended ranges are shown, but analysis was mostly developed on actual occurrence data

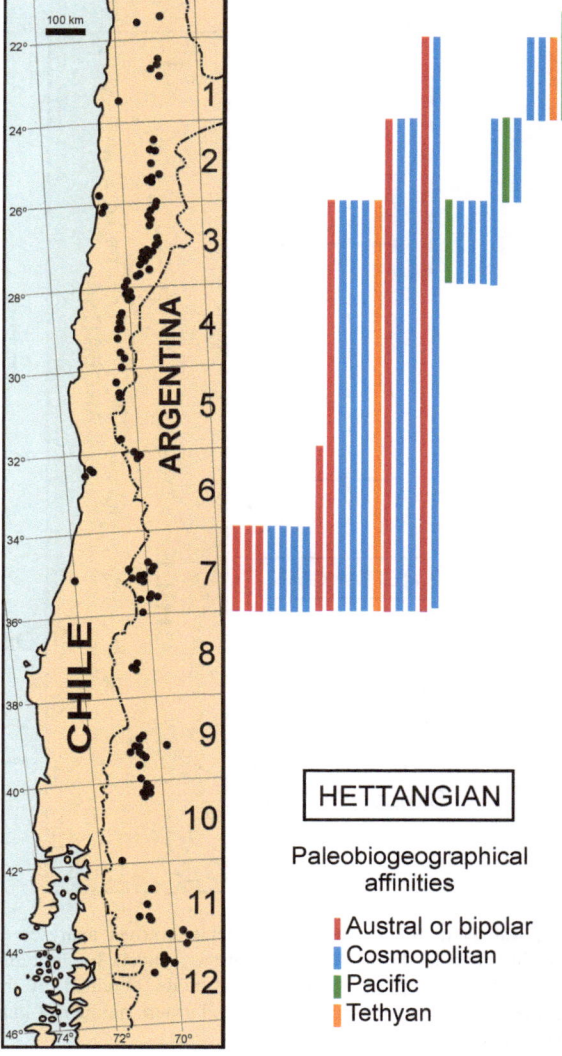

On the other hand, species with warm-water affinities are few: *Eopecten velatus* (Goldfuss) and a *Parallelodon* species.

With these data, it is evident that high-latitude species are relatively abundant south to 32° but extend their influence at least to 22°. This is here regarded as the range of mixed faunas between the Austral and Tethyan Realms for Hettangian times (about 900 km wide).

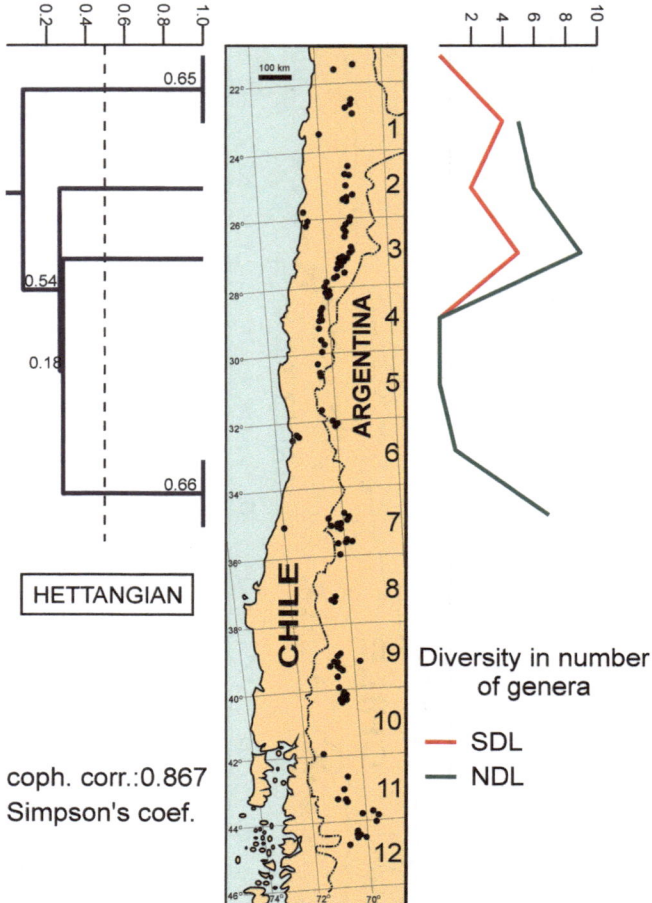

Fig. 4.8 Hettangian: cluster and faunal turnover analyses along the studied latitudinal range, locality map in the center. *Left* Cluster analysis for latitudinal gradient, values on each node represent the support value for the node obtained by bootstrapping. Similarity measure: Simpson's coefficient; algorithm: paired group; number of iterations for the bootstrapping: 1,000. *Right* Faunal turnover analysis

4.2.2.2 Sinemurian

During the Sinemurian (Fig. 4.9) there seems to be a southwards shift in the main turnover region, as indicated by the cluster analysis (grouping the zones between 26° and 32° S on one hand, and those between 32° and 36° S on the other, Fig. 4.10) as well as by the limits of distribution (showing a peak of northern and southern limits between 28° and 30° S). The minor inconsistency between both types of data may be due to scarcity of records in some regions, being more reliable the limit suggested by the faunal turnover.

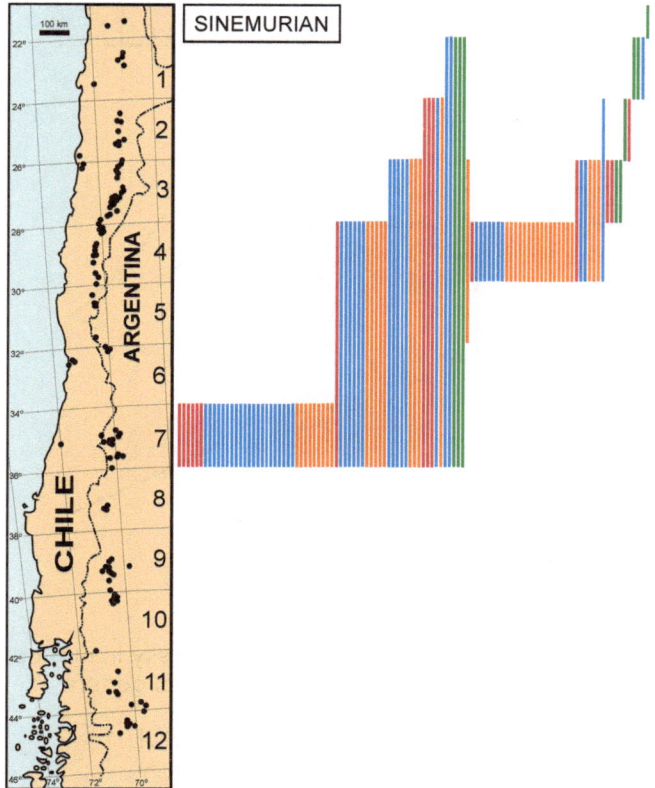

Fig. 4.9 Latitudinal ranges of Sinemurian bivalve species (each vertical line represents one species) discriminated by paleobiogeographic affinities (line key in Fig. 4.7). The 13 latitudinal areas used for this analysis (spanning 2° each) are numbered, and each locality is represented by a *black dot*. Notice that localities south of 36° do not bear Sinemurian faunas. Extended ranges are shown, but analysis was mostly developed on actual occurrence data

Once again there is a strong southwards reduction (Fig. 4.15b) in the relative number of species of Pacific affinities (linear predictor: −0.23; $p < 0.001$), while cosmopolitan species seem to increase in that direction (linear predictor: 0.08; $p = 0.045$).

Species with high-latitude affinities (austral or bipolar) which reach the northernmost regions (up to 24°–26°) are: *Agerchlamys*? sp., *Otapiria pacifica*, *O. neuquensis* Damborenea, *Kalentera* sp., *Parainoceramus apollo*? (Leanza), and *Harpax rapa* (Bayle and Coquand). The southern diversity peak for high-latitude taxa also contains some of these species, but to these, species of Astartidae, *Asoella*, and *Kolymonectes* are added. On the other hand, the number of species with warm-water affinities is reduced south of 32° (the southernmost increase is due to the occurrence of just one isolated species in that region).

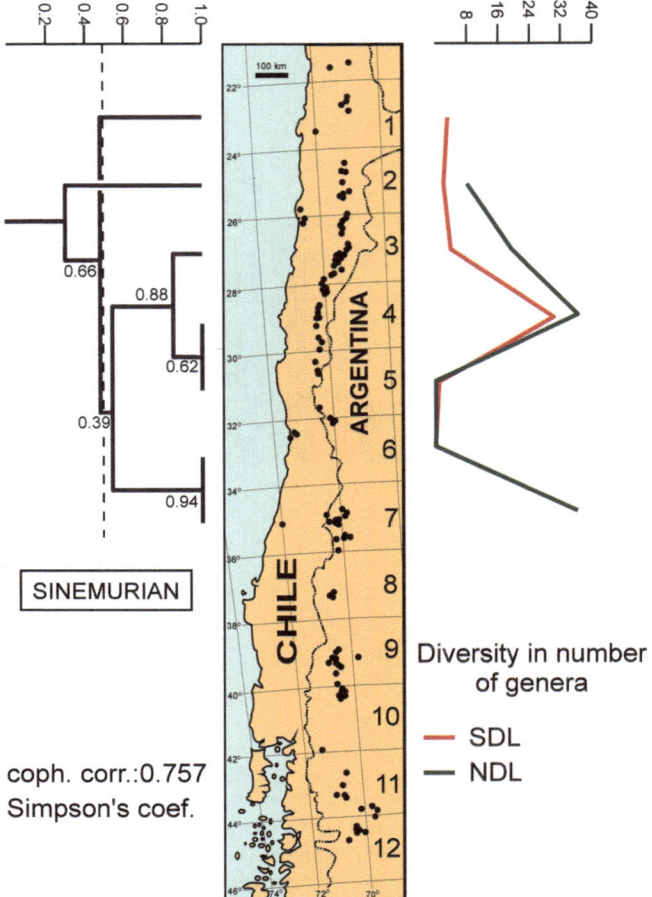

Fig. 4.10 Sinemurian: cluster and faunal turnover analyses along the studied latitudinal range, locality map in the center. *Left* Cluster analysis for latitudinal gradient, values on each node represent the support value for the node obtained by bootstrapping. Similarity measure: Simpson's coefficient; algorithm: paired group; number of iterations for the bootstrapping: 1,000. The dendrogram was drawn in three dimensions to fit it to the geographic relative positions of the analyzed areas. *Right* Faunal turnover analysis

4.2.2.3 Pliensbachian

Reliable data to the north of 26° are very scarce and have only been included for the sake of completeness. Otherwise, bivalve faunas of this age are by far the best known for the Early Jurassic of the southern Andean region (Fig. 4.11).

As already said, bivalve faunas show a sharp rise in overall diversity during the Pliensbachian (see Fig. 4.6), which may be only partially attributed to intensity of studies. All elements of the fauna participate in this increase in species numbers.

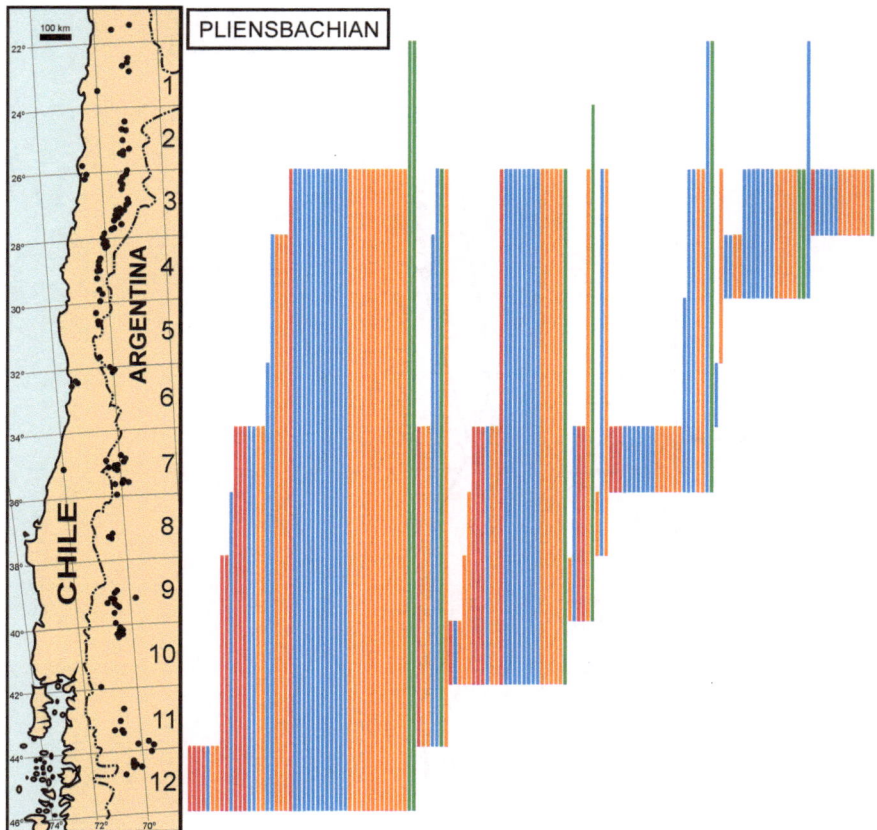

Fig. 4.11 Latitudinal ranges of Pliensbachian bivalve species (each vertical line represents one species) discriminated by paleobiogeographic affinities (line key in Fig. 4.7). The 13 latitudinal areas used for this analysis (spanning 2° each) are numbered, and each locality is represented by a black dot. Notice that localities north of 26° have very few data. Extended ranges are shown, but analysis was mostly developed on actual occurrence data

Along the whole study area, bivalves with Tethyan affinities maintain a steady percentage and coexist with high-latitude taxa.

The complete Pliensbachian database allowed for the most detailed analysis of the paleobiogeography of the west margin of southern South America. Cluster analysis (Fig. 4.12) allowed for the discrimination of northern latitudes (22°–32° S) from southern ones (32°–46°), although the best defined biogeographic region is between 34° and 44° S (i.e., coinciding with the Neuquén embayment at the time). According to the limits of distribution (Fig. 4.12), and in coincidence with the cluster analysis, the main biogeographic turnover seemed to be between 34° and 36° S during this stage, showing an even greater displacement toward south. It is noteworthy that the

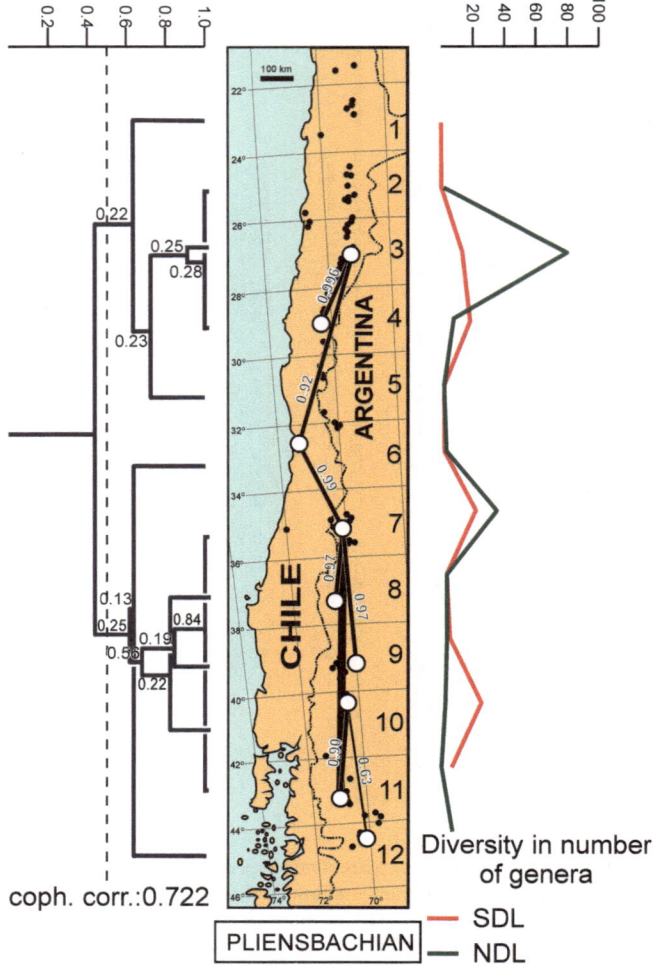

Fig. 4.12 Pliensbachian: cluster and faunal turnover analyses along the studied latitudinal range, locality map in the center. *Left* Cluster analysis for latitudinal gradient, values on each node represent the support value for the node obtained by bootstrapping. Similarity measure: Simpson's coefficient; algorithm: paired group; number of iterations for the bootstrapping: 1,000. The dendrogram was drawn in three dimensions to fit it to the geographic relative positions of the analyzed areas. *Right* Faunal turnover analysis. *On the map* BSN analysis; edge thickness indicating dissimilarity: thick line: dissimilarity lower than 0.1, intermediate line: dissimilarity equal or higher than 0.1 but lower than 0.25, thin line: dissimilarity equal or higher than 0.25 but lower than 0.5; values on the edges: support value for each edge (value is 1 when there is no indication)

peak of southern limits of distribution is strongly affected by species of Tethyan affinities (8), while the one of northern limits is affected by species with high-latitude affinities (12) and also by species of Tethyan affinities (12). The main change at this boundary is marked by a clear increment toward the south (linear predictor: 0.12; $p \ll 0.001$) in the proportion of species with high-latitude affinities (either austral or bipolar) associated with a reduction (linear predictor: -0.03; $p = 0.047$) in cosmopolitan species (Fig. 4.15c).

For this stage a BSN was constructed (Fig. 4.12), obtaining similar results; a southern region can be recognized, mostly due to the similarity of many latitudinal bins with that between 34° and 36° S. This points once again to a biogeographic turnover at that region. Also, at this stage a significant positive correlation was found between dissimilarity value and latitudinal separation among samples (Spearman's $r_s = 0.61$, $p \ll 0.01$) indicating a clear latitudinal gradient.

High-latitude taxa are relatively abundant south of 34° (Fig. 4.15c), but *Harpax rapa*, *Radulonectites sosneadoensis* (Weaver), and a species of *Palaeopharus*?, all of them with bipolar affinities, reach the 26–28° region. High-latitude faunas south of 34° are more varied, including 12 species south of 40°. The more conspicuous elements of this Austral fauna are: *Parainoceramus apollo* (Leanza)*, Kolymonectes weaveri* Damborenea, *Agerchlamys wunschae* (Marwick), *Otapiria neuquensis* Damborenea, and *Kalentera riccardii* Damborenea.

4.2.2.4 Toarcian

Toarcian faunas, though widespread and relatively abundant, are less well known than Pliensbachian ones, especially south of 40° (Fig. 4.13). Consequently, the results for the Toarcian seem a little obscure, at least for the ordination methods. The cluster analysis (Fig. 4.14) shows no clear pattern, while on the graphics for the limits of distribution there are several peaks. Nevertheless, some significant trends were recognized on the proportional representation of species with different biogeographic affinities (Fig. 4.15d). There is a significant increase in high-latitude species (both austral and bipolar) toward the south (linear predictor: 0.21; $p = 0.001$), and probably also in Pacific species (linear predictor: 0.05; $p = 0.092$); on the other hand, species of Tethyan affinities significantly decrease in relative number in that same direction (linear predictor: -0.06; $p = 0.033$). These changes seem to be gradual, occurring between 32° and 42° S.

There is a decrease in the total number of taxa with high-latitude (bipolar + austral) affinities, from 7 in the Hettangian, 15 in the Sinemurian, and 22 in the Pliensbachian, to only 5 species in the Toarcian (Fig. 4.6). By Toarcian times, true high-latitude taxa, such as *Harpax rapa*, *Arctotis? frenguellii* Damborenea, an austral species of *Entolium* (*E. mapuche* Damborenea), and a species of Inoceramidae, do not extend north of 34°, and only *Meleagrinella*, which has bipolar affinities but is nevertheless rather cosmopolitan in distribution, reaches 32°. On the other hand, several elements of the *Lithiotis* reef facies occur in the northern part of the range, including *Lithiotis* down to the 28°–30° region. *Gervilleioperna* and *Opisoma* extend even further south.

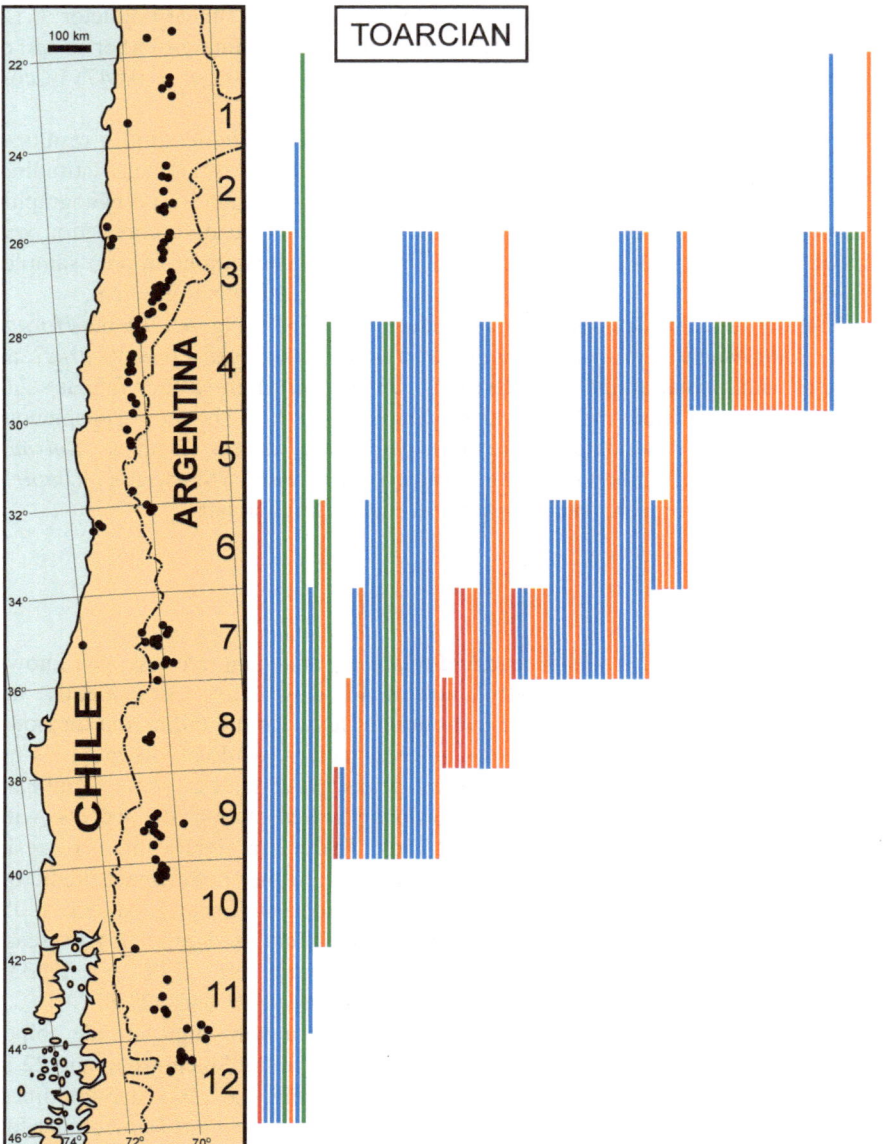

Fig. 4.13 Latitudinal ranges of Toarcian bivalve species (each vertical line represents one species), discriminated by paleobiogeographic affinities (line key in Fig. 4.7). The 13 latitudinal areas used for this analysis (spanning 2° each) are numbered, and each locality is represented by a *black dot*. Notice that localities north of 26° and south of 40° have very few data. Extended ranges are shown, but analysis was mostly developed on actual occurrence data

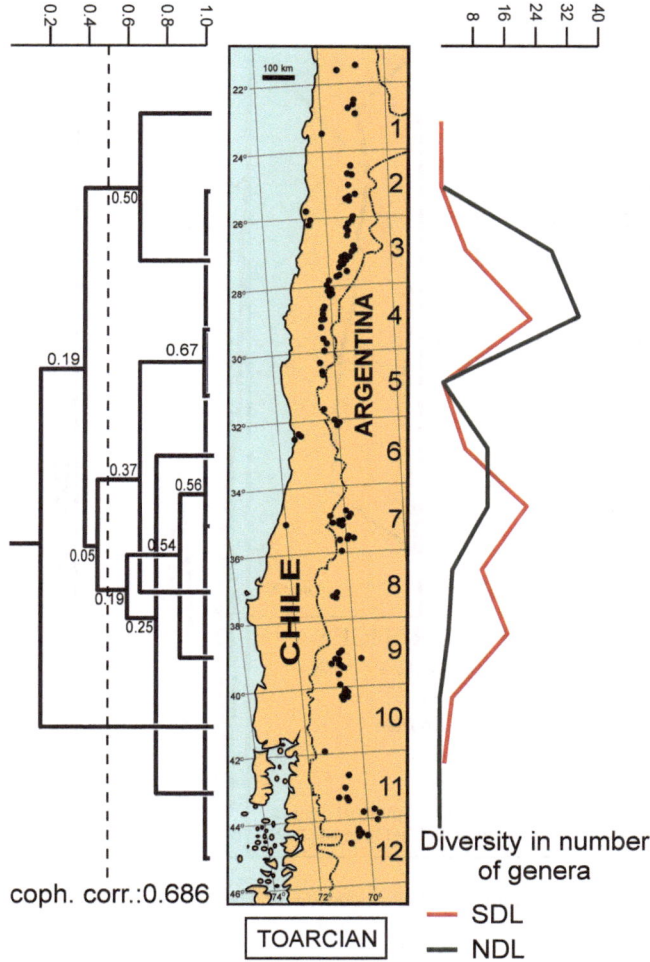

Fig. 4.14 Toarcian: cluster and faunal turnover analyses along the studied latitudinal range, locality map in the center. *Left* Cluster analysis for latitudinal gradient, values on each node represent the support value for the node obtained by bootstrapping. Similarity measure: Simpson's coefficient; algorithm: paired group; number of iterations for the bootstrapping: 1,000. The dendrogram was drawn in three dimensions to fit it to the geographic relative positions of the analyzed areas. *Right* Faunal turnover analysis

In summary, this brief account of the temporal changes shown by the latitudinal distribution of bivalves indicates that both the faunal turnover analysis (Figs. 4.8, 4.10, 4.12, and 4.14) and the proportion of species with high-latitude biogeographic affinities (Fig. 4.15) consistently confirm a southward migration of the boundary between northern and southern faunas through time (from Hettangian to Toarcian). This will be discussed in detail in Sect. 4.3, see also Table 4.3.

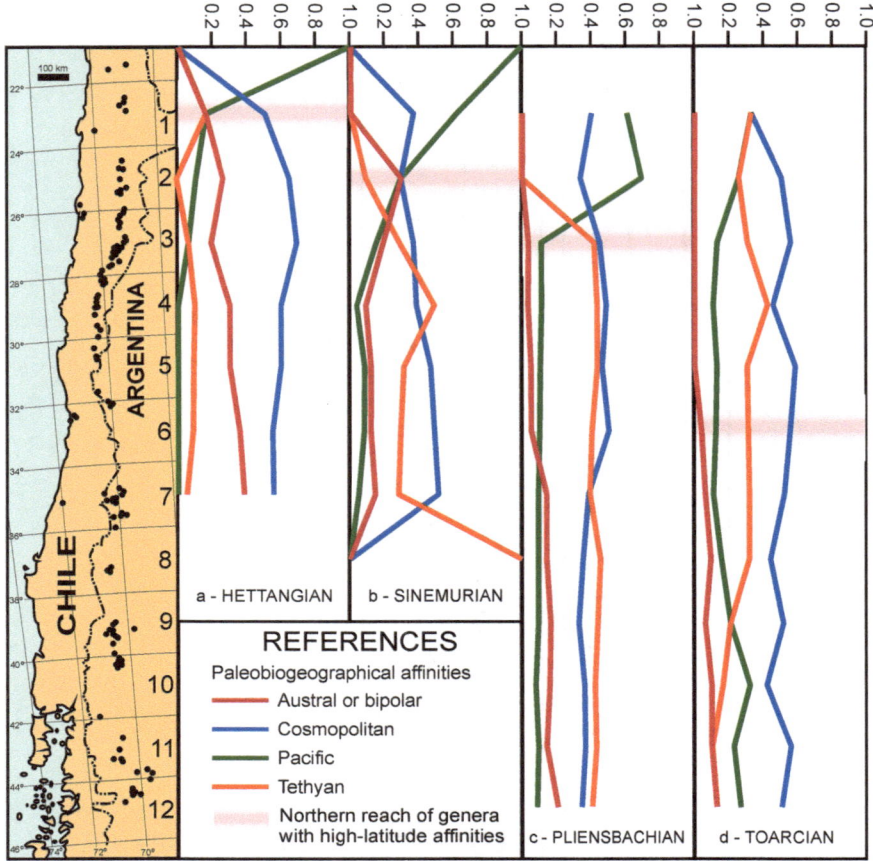

Fig. 4.15 Latitudinal variations in the proportion of species discriminated by their biogeographic affinities for the four time intervals analyzed. The northernmost reach of species with high-latitude affinities is also shown, and shows a steady migration toward the south with time

Table 4.3 Comparison of the results obtained using three different methods to recognize the latitudinal position of the boundary between Tethyan and South Pacific faunas for each of the Early Jurassic stages

	Cluster analysis (here)	Faunal turnover analysis (here)	Northernmost reach of species with high-latitude affinities
HETTANGIAN	30°	26–28°	22–24°
SINEMURIAN	32°	28–30°	24–26°
PLIENSBACHIAN	32°	34–36°	26–28°
TOARCIAN	?	?	32–34°

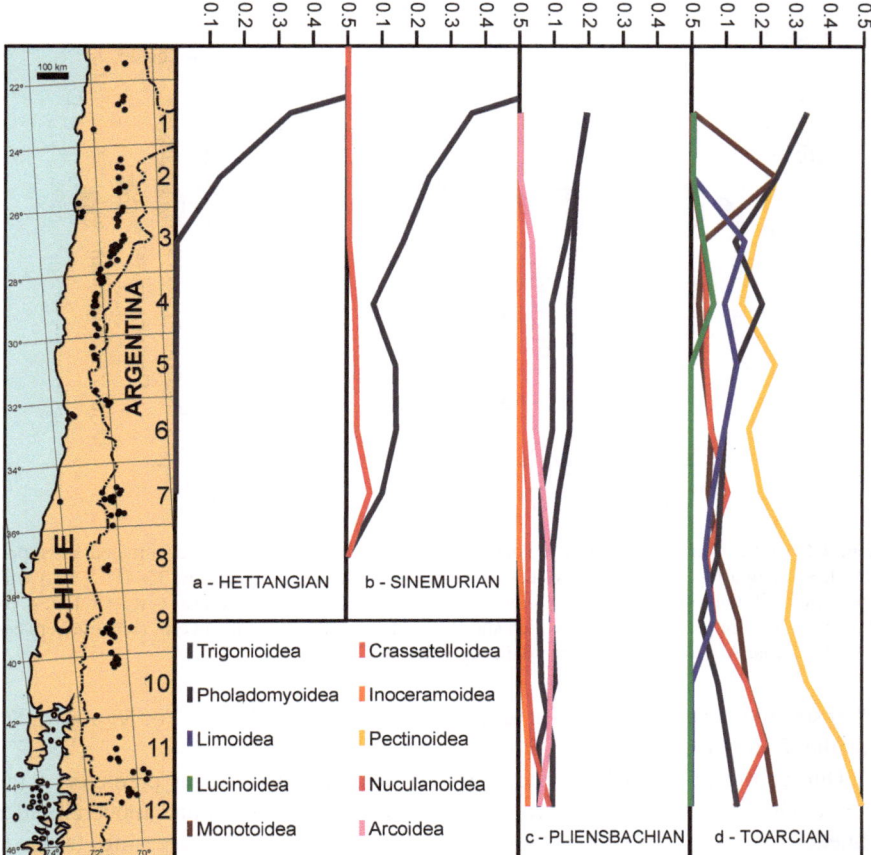

Fig. 4.16 Latitudinal variations in the proportion of species among superfamilies; only those superfamilies with a significant variation for each stage are shown (explanation in the text)

4.2.3 Systematic Latitudinal Variations

The analysis for changes in proportion of different systematic groups was applied to the superfamilies represented on each stage (Fig. 4.16), but in most cases, due to the low number of species in each group, there were no significant results, especially for the Hettangian and Sinemurian. Nevertheless, it is noteworthy the trend of decreasing relative diversity at higher latitudes for the superfamily Trigonioidea in the four considered stages (Fig. 4.16) only when range extension was applied (linear predictor: -1.09; $p = 0.025$ for the Hettangian; linear predictor: -0.10; $p = 0.067$ for the Sinemurian; linear predictor: -0.06; $p = 0.033$ for the

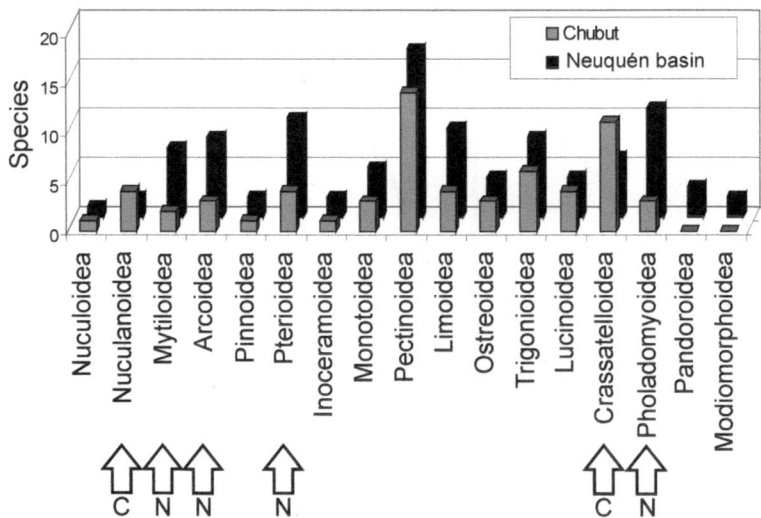

Fig. 4.17 Late Pliensbachian/Early Toarcian bivalve species richness discriminated by super-families, comparing faunas from Mendoza-Neuquén provinces (present-day 32–41° S lat.) with those from Chubut province (present-day 42–45° S lat.), see location of these regions in Fig. 4.1b. The *arrows* point to superfamilies which show significant differences in species richness between the two areas: *N* richer in Neuquén basin, *C* richer in Chubut

Pliensbachian excluding interval areas 1 and 2; linear predictor: -0.09; $p = 0.026$ for the Toarcian).

During the Pliensbachian, the superfamilies Arcoidea (linear predictor: 0.06; $p = 0.036$), Nuculanoidea (linear predictor: 0.10; $p = 0.025$) and Inoceramoidea (linear predictor: 0.30; $p = 0.042$) showed a southward increasing trend in proportion of species, while Pholadomyoidea (linear predictor: -0.05; $p = 0.008$) and possibly Trigonioidea showed an opposite trend. The low values on the proportion of species of the different families are remarkable for this stage (Fig. 4.16), and they were not restricted only to the mentioned superfamilies. For the Toarcian the superfamilies Limoidea (linear predictor: -0.09; $p = 0.048$), Lucinoidea (linear predictor: -0.31; $p = 0.027$), and Trigonioidea decreased southwards in relative number of species, while Monotoidea (linear predictor: 0.15; $p = 0.001$), Pectinoidea (linear predictor: 0.07; $p = 0.009$) and Crassat-elloidea (linear predictor: 0.09; $p = 0.033$) tend to increase their relative diversity in that same direction (Fig. 4.16); superfamily Crassatelloidea showed the same trend during the Sinemurian (linear predictor: 0.41; $p = 0.048$) despite its low overall diversity.

At a regional scale in western Argentina, a general comparison of the late Pliensbachian-early Toarcian faunas from Mendoza/Neuquén and those from Chubut (Damborenea et al. 2010) done as part of an ongoing detailed systematic revision of the latter shows that species richness within most taxonomic groups was almost equal in both areas (pectinoids and trigonioids, for example, to name diverse Jurassic groups). On the other hand, it is also evident that a few clades had notable differences in diversity between both areas (Fig. 4.17). Some superfamilies were significantly more diverse in Chubut (such as nuculanoids and crassatelloids), while several others (for instance mytiloids, arcoids, pterioids and pholadomyoids) were more diverse in the northern region. Due to the limited time span of the Chubut extensive marine deposits, these trends can only be noticed at this particular time slice.

Knowledge on the latitudinal distribution of living bivalves (Crame 2000a, b, 2001, 2002) just shows that some of the mentioned superfamilies are nowadays latitudinally limited in their distribution, or have very steep diversity gradients toward the poles. Extant mytiloids, arcoids, pholadomyoids, and pterioids have a steep decrease toward high latitudes, while protobranchs show no significant latitudinal gradient. The observed pattern in our data suggest that this particular trends may be considerably older than previously thought for some groups, and it is difficult to understand them as being originated solely by clade age as postulated by Crame. It is also interesting to note that the Jurassic was characterized by temperature gradients less evident than at the present (see Sect. 1.5), and even so these selective diversity gradients are revealed.

4.3 Distribution Patterns and Boundaries

Throughout the Jurassic, as just seen, the boundary zone between South Pacific and Tethyan paleobiogeographic units fluctuated in position through time (Damborenea 2002b). The approximate latitudinal location of the transitional boundary area and its shift through time were long ago recognized on the basis of faunal composition along the Andean region (Damborenea 1996, 2002b). Those early studies are now confirmed by the quantitative analysis presented here: along the South American western margin, the faunal change boundary shifted southwards between 8 and 10 latitude degrees, and this is consistently indicated (though figures obtained are not strictly coincident) by the three methods used so far (Table 4.3).

It is well known that biogeographic boundaries are not sharp borders but transitional zones between different faunas, and therefore it is difficult to determine them precisely. These transitional zones have also been reported and studied in the Early Jurassic deposits of the Northern Hemisphere based both on bivalves (for instance Hayami 1990) or ammonites (see Dommergues and Meister 1991, with further references), and will be discussed later (Sect. 5.4).

As already mentioned, the results presented here agree to indicate that the boundary zone shifted latitudinally southwards during the Early Jurassic (Table 4.3). This observation holds true either if the northermost boundary of this mixed-fauna fringe is taken into account ($26° \rightarrow 28° \rightarrow 34°$ for the turnover analysis, or $22° \rightarrow 24° \rightarrow 26° \rightarrow 32°$ for the high-latitude species reach) or the southernmost limit is considered ($28° \rightarrow 30° \rightarrow 36°$ for the turnover analysis, or $24° \rightarrow 26° \rightarrow 28° \rightarrow 34°$ for the high-latitude species reach). This southwards shift in the boundary between the Austral and Tethyan Realms amounts, therefore, to about 700 km (about 8°) along the margin of the Jurassic Pacific.

This observed shift (recorded on present-day maps) of an Early Jurassic paleobiogeographic boundary in this area of the Pacific could be interpreted as a result of either: (a) drifting of continental masses without significant changes in climatic conditions, (b) changing of overall climatic conditions without significant latitudinal continental drift, or, more likely, (c) a combination of factors involving both climatic change and continental latitudinal drift. At the present state of knowledge, it is very difficult to identify and test separately the two main possible causes. Pole paleopositions based on paleomagnetic studies and water paleotemperatures inferred from isotopic distributions are instrumental to this discussion.

Paleomagnetic data (Iglesia-Llanos et al. 2006; Iglesia-Llanos 2012) and "absolute" paleogeographic reconstructions of Pangea for the Early Jurassic suggest that it moved northwards, attaining its northernmost position by Late Pliensbachian-Early Toarcian, and shifted several latitude degrees from Hettangian to Toarcian times. Despite the paucity of available paleomagnetic data, it is evident that the results presented here are broadly consistent with them. This is in general agreement with the observed shift in the bivalve distribution boundaries, and will be treated in more detail in a global-scale discussion (Sect. 5.4).

References

Aberhan M (1992) Palöokologie und zeitliche Verbreitung bentischer Faunengemeinschaften im Unterjura von Chile. Beringeria 5:1–174

Aberhan M (1993) Benthic macroinvertebrate associations on a carbonate-clastic ramp in segments of the Early Jurassic back-arc basin of northern Chile (26-29°S). Rev Geol Chile 20:105–136

Aberhan M (1994) Early Jurassic Bivalvia of northern Chile. Part I. Subclasses Palaeotaxodonta, Pteriomorphia, and Isofilibranchia. Beringeria 13:1–115

Aberhan M (2004) Early Jurassic bivalvia of northern Chile. Part II: Subclass Anomalodesmata. Beringeria 34:117–154

Aberhan M, Hillebrandt A (1996) Taxonomy, ecology, and palaeobiogeography of *Gervilleioperna (Gervilleiognoma) aurita* n. subgen. n. sp. (Bivalvia) from the Middle Jurassic of northern Chile. Paläontol Z 70:79–96

Aberhan M, Hillebrandt A (1999) The bivalve *Opisoma* in the lower Jurassic of northern Chile. Profil 16:149–164

Aberhan M, Pálfy J (1996) A low oxygen tolerant East Pacific flat clam (*Posidonotis semiplicata*) from the lower Jurassic of the Canadian cordillera. Canad J Earth Sci 33:993–1006

Al-Suwaidi AH, Angelozzi GN, Baudin F, Damborenea SE, Hesselbo SP, Jenkyns HC, Manceñido MO, Riccardi AC (2010) First record of the early Toarcian Oceanic anoxic event from the Southern hemisphere, Neuquén basin, Argentina. J Geol Soc Lon 167:633–636

Bayle E, Coquand H (1851) Mémoire sur les Fossiles recueillis dans le Chili par M. Ignace Domeyko et sur les terrains auxquels ils appartiennent. Mém Soc Géol France ser. 2, 4:1–47

Behrendsen O (1891) Zur Geologie des Ostabhanges der argentinischen Cordillere. Teil I. Z Deuts Geolog Gesells 43:369–420

Benito J, Chernicoff J (1986) Geología del Cerro Caquel y aledaños. Departamento Futaleufú. Provincia del Chubut. Rev Asoc Geol Argent 41:70–80

Blasco G, Levy R, Plozkiewicz V (1980) Las calizas toarcianas de Loncopán, Departamento Tehuelches, Provincia del Chubut, República Argentina. Actas 2° Congr Argent Paleontol Bioestratigr 1:191–200

Burmeister H, Giebel C (1861) Die Versteinerungen von Juntas im Thal des Rio de Copiapó. Abh Naturfors Gesells Halle 6:1–34

Cecioni G, Westermann GEG (1968) The Triassic/Jurassic marine transition of coastal Central Chile. Pacific Geol 1968, 1:41–75

Chong DG, Hillebrandt A (1985) El Triásico preandino de Chile entre los 23° 30' y 26° 00' de lat. Sur. Actas 4° Congr Geol Chileno (Antofagasta) 1:162–210

Conrad TA (1855) Remarks on the fossil shells from Chile, collected by Lieut. Gillis, with description of the species. US Naval Astronom Exped Southern Hemisphere 1849-50-52. 2 H (Paleontol):282–286

Cortiñas JS (1984) Estratigrafía y facies del Jurásico entre Nueva Lubecka, Ferrarotti y Cerro Colorado. Su relación con los depósitos coetáneos del Chubut central. Actas 9° Congr Geol Argent 2:283–299

Covacevich V, Escobar F (1979) La presencia del género *Otapiria* Marwick, 1935 (Mollusca: Bivalvia) en Chile y su distribución en el ámbito circumpacífico. 2° Congr Geol Chileno:H165–H187

Crame JA (2000a) Evolution of taxonomic diversity gradients in the marine realm: evidence from the composition of recent bivalve faunas. Paleobiology 26:188–214

Crame JA (2000b) The nature and origin of taxonomic diversity gradients in marine bivalves. In: Harper EM, Taylor JD, Crame JA (eds) The evolutionary biology of the Bivalvia. Geol Soc Spec Publ 177:347–360

Crame JA (2001) Taxonomic diversity gradients through geological time. Divers Distrib 7:175–189

Crame JA (2002) Evolution of taxonomic diversity gradients in the marine realm: a comparison of late Jurassic and recent bivalve faunas. Paleobiology 28:184–207

Damborenea SE (1987a) Early Jurassic Bivalvia of Argentina. Part I: Stratigraphical Introduction and Superfamilies Nuculanacea, Arcacea, Mytilacea and Pinnacea. Palaeontographica A 99:23–111, pl 1–4

Damborenea SE (1987b) Early Jurassic Bivalvia of Argentina. Part 2: Superfamilies Pteriacea, Buchiacea and part of Pectinacea. Palaeontographica A 199:113–216, pl 1–14

Damborenea SE (1989) El género *Posidonotis* Losacco (Bivalvia, Jurásico inferior): su distribución estratigráfica y paleogeográfica. Actas 4° Congr Argent Paleontol y Bioestratigr (Mendoza 1986) 4:45–51

Damborenea SE (1993) Early Jurassic South American pectinaceans and circum-Pacific palaeobiogeography. Palaeogeogr Palaeoclimatol Palaeoecol 100:109–123

Damborenea SE (1996) Palaeobiogeography of early Jurassic bivalves along the southeastern Pacific margin. 13° Congr Geol Argent y 3° Congr Explorac Hidrocarb (Buenos Aires) Actas 5:151–167

Damborenea SE (1998) The bipolar bivalve *Kolymonectes* in South America and the diversity of Propeamussiidae in Mesozoic times. In: Johnston PA, Haggart JW (eds) Bivalves: an eon of evolution—paleobiological studies honoring Norman D. Newell University of Calgary Press, Calgary

Damborenea SE (2002a) Early Jurassic bivalves from Argentina. Part 3: Superfamilies Monotoidea, Pectinoidea, Plicatuloidea and Dimyoidea. Palaeontograph A 265:1–119

Damborenea SE (2002b) Jurassic evolution of Southern Hemisphere marine palaeobiogeographic units based on benthonic bivalves. Geobios 35(MS (24)):51–71

Damborenea SE (2004) Early Jurassic *Kalentera* (Bivalvia) from Argentina and its palaeobiogeograhical significance. Ameghiniana 41:185–198

Damborenea SE, Lanés S (2007) Early Jurassic shell beds from marginal marine environments in southern Mendoza, Argentina. Palaeogeogr Palaeoclimatol Palaeoecol 250:68–88

Damborenea SE, Manceñido MO (2012) Late Triassic bivalves and brachiopods from southern Mendoza, Argentina. Rev Paléobiol, VS 11:317–344

Damborenea SE, Pagani MA, Ferrari SM (2010) Paleogeografía del Jurásico temprano de Chubut: aportes de los moluscos. Res 4° Simp Argent Jurásico, Bahía Blanca: 35

Dediós P (1967) Geología del cuadrángulo Vicuña, Provincia de Coquimbo. Ins Inv Geol Chile Carta 16:1–85

Dommergues JL, Meister C (1991) Area of mixed marine faunas between two major paleogeographical realms, exemplified by the early Jurassic (Late Sinemurian and Pliensbachian) ammonites in the Alps. Palaeogeogr Palaeoclimatol Palaeoecol 86:265–282

Feruglio E (1934) Fossili Liassici della Valle del Rio Genua (Patagonia). Gior Geol Ann R Mus Geol Bologna 9:1–64

Goodwin DH (1997) The importance of paleoecology in assessing paleogeographic relationships of the Antimonio Formation. In: González-León CM, Stanley GD jr (eds) US-Mexico cooperative research. International workshop on the geology of Sonora Memoir. Univ Nac Autón México, Inst Geol, Estación Regional del Noroeste, Publ Ocas, vol 1, pp 36–38

Groeber P (1946) Observaciones geológicas a lo largo del meridiano 70. Hoja Chos Malal. Rev Soc Geol Argent 1:177–206

Groeber P, Stipanicic P, Mingramm A (1953) Jurásico. In: Geografía de la República Argentina II (1a. parte): Mesozoico. Soc Argent Estud Geogr GAEA:143–347

Gulisano C (1992) Paleogeographical evolution of west-central Argentina. In: Westermann GEG (ed) The Jurassic of the Circum-Pacific. Cambridge University Press, London

Hallam A (1977) Jurassic bivalve biogeography. Paleobiology 3:58–73

Harrington HJ (1961) Geology of parts of Antofagasta and Atacama Provinces, northern Chile. Bull Am Assoc Petrol Geol 45:169–197

Hayami I (1988) A Tethyan bivalve, *Posidonotis dainellii*, from the lower Jurassic of West Japan. Trans Proc Palaeontol Soc Jpn NS 151:564–569

Hayami I (1990) Geographic distribution of Jurassic faunas in Eastern Asia. In: Ichikawa K, Mizutani S, Hara I, Hada S, Yao A (eds) Pre-Cretaceous terranes of Japan. Publication of IGCP project 224, Osaka

Hillebrandt A (1971) Der Jura in der chilenisch-argentinischen Hoch cordillere (32° 30 S). Münster Forsch Geol Paläontol 20/21:63–87

Hillebrandt A (1973) Neue Ergebnisse über den Jura in Chile und Argentinien. Münster Forsch Geol Paläontol 31/32:167–199

Hillebrandt A (1977) Ammoniten aus dem Bajocien (Jura) von Chile (Südamerika). Neue Arten der Gattungen *Stephanoceras* und *Domeykoceras* n. gen. (Stephanoceratidae). M. Mitt Bayer Staats Paläontol Hist Geol 17:35–69

Hillebrandt A (1980) Paleozoogeografía de Jurásico marino (Lías hasta Oxfordiano) en Suramérica. In: Zeil W (ed) Nuevos resultados de la investigación geocientífica alemana en Latinoamérica. Deuts Forschungs and Inst Colabor Cient, Tübingen:123–134

Hillebrandt A (1990) Der Untere Jura im Gebiet des Rio Atuel (Provinz Mendoza, Argentinien). N Jb Geol Paläontol Abh 181:143–157

Hillebrandt A (2000) Ammonite biostratigraphy of the Hettangian/Sinemurian boundary in South America. In: Hall RL, Smith PL (eds) Advances in Jurassic Research 2000. GeoRes Forum 6:105–118

Hillebrandt A (2002) Ammoniten aus dem oberen Sinemurium von Südamerika. Revue Paléolbiol 21:35–147

Hillebrandt A, Schmidt-Effing R (1981) Ammoniten aus dem Toarcium (Jura) von Chile (Südamerika). Die Arten der Gattungen *Dactylioceras*, *Nodicoeloceras*, *Peronoceras* und *Collina*. Zitteliana 6:1–74

Hillebrandt A, Westermann GEG (1985) Aalenian (Jurassic) ammonite faunas and zones of the southern Andes. Zitteliana 12:3–55

Hillebrandt A, Gröschke M, Prinz P, Wilke HG (1986) Marienes Mesozoikum in Nordchile zwischen 21° und 26°S. Berliner Geowiss Abh A 169:1–40

Iglesia-Llanos MP (2012) Palaeomagnetic study of the Jurassic from Argentina: magnetostratigraphy and palaeogeography of South America. Rev Paléobiol V S 11:151–168

Iglesia-Llanos MP, Riccardi AC, Singer SE (2006) Palaeomagnetic study of lower Jurassic marine strata from the Neuquén Basin, Argentina: a new Jurassic apparent polar wander path for South America. Earth Planet Sci Let 252:379–397

Jablonski D, Roy K, Valentine JW (2000) Analysing the latitudinal gradient in marine bivalves. In: Harper EM, Taylor JD, Crame JA (eds) The evolutionary biology of the Bivalvia. Geol Soc Spec Publ 177:361–365

Jaworski E (1925) Contribución a la paleontología del Jurásico Sudamericano. Publ Dir Gen Min Geol Hidrol sec Geol 4:1–160

Kindlmann P, Schödelbauerová I, Dixon AFG (2007) Inverse latitudinal gradients in species diversity. In: Storch D, Marquet PA, Brown JH (eds) Scaling Biodiversity. Cambridge University Press, Cambridge

Krug AZ, Jablonski D, Valentine JW (2007) Contrarian clade confirms the ubiquity of spatial origination patterns in the production of latitudinal diversity gradients. Proc Nat Acad Sci USA 104:18129–18134

Lage J (1982) Descripción geológica de la Hoja 43C Gualjaina, Prov. del Chubut. Bol Serv Geol Nac 189:1–72

Leanza AF (1942) Los pelecípodos del Lías de Piedra Pintada, en el Neuquén. Rev Mus La Plata (ns) Paleontol 2:143–206

Legarreta L, Uliana MA (1996) The Jurassic succession in west-central Argentina: stratal patterns, sequences and paleogeographic evolution. Palaeogeogr Palaeoclimatol Palaeoecol 120:303–330

Legarreta L, Uliana MA (2000) El Jurásico y Cretácico de la Cordillera Principal y la Cuenca Neuquina. 1, Facies sedimentarias. Inst Geol Rec Mineral 29:399–432

Lesta P, Ferello R, Chebli G (1980) Chubut Extraandino. In: Turner J (ed), Segundo Simposio de Geología Regional Argentina II, Acad Nac Cienc (Argent):1307–1387

Malumián N, Ploszkiewicz VA (1976) El Liásico fosilífero de Loncopán, Departamento Tehuelches, Provincia de Chubut. Rev Asoc Geol Argent 31:279–280

Manceñido MO, Damborenea SE (1984) Megafauna de Invertebrados paleozoicos y mesozoicos. In: Ramos VA (ed), Geología y recursos naturales de la Provincia de Rio Negro. Relatorio 9° Congr Geol Argent:413–465

Massaferro GI (2001) El Jurásico temprano del cerro Cuche (Cordillera Patagónica del Chubut): estratigrafía y fósiles. Rev Asoc Geol Argent 56(2):244–248

Mercado M (1982) Hoja Laguna del Negro Francisco, Región de Atacama. Serv Nac Geol Min, Carta Geol Chile 56:1–63

Möricke W (1894) Versteinerungen des Lias und Unteroolith von Chile. N Jb Min Geol Paläontol BB 9:1–100

Mpodozis C, Rivano S, Vicente JC (1973) Resultados preliminares del estudio geológico de la Alta Cordillera de Ovalle entre los ríos Grande y Los Molles (Prov de Coquimbo, Chile). Actas 5° Congr Geol Argent: 117–132

Nullo FE (1983) Descripción Geológica de la Hoja 45c, Pampa de Agnia, Provincia de Chubut. Bol Serv Geol Nac 199:1–94

Pagani MA, Manceñido MO, Damborenea SE, Ferrari SM (2012) The ichnogenus *Lapispira* from the early jurassic of patagonia (Chubut, Argentina). Rev Paléobiol VS 11: 409–416

Pérez E, Levi B (1961) Relación estratigráfica entre la Formación Montezunia y el granito subyacente, Calama, Prov. de Antofagasta. Chile. Rev Min 16:39–48

Pérez E, Reyes R (1977) Las trigonias jurásicas de Chile y su valor cronoestratigráfico. Bol Inst Investig Geol Chile 30:1–58

Pérez E, Reyes R (1994) Catálogo de ejemplares tipo, conservados en la Colección Paleontológica del Servicio Nacional de Geología y Minería, Chile. Serv Nac Geol Min Bol 46:1–99

Pérez E, Reyes R, Damborenea SE (1995) El género *Groeberella* Leanza, 1993 y Groeberellidae nov. (Bivalvia; Trigonioida) del Jurásico de Chile y Argentina. Rev Geol Chile 22:143–157

Pérez E, Aberhan M, Reyes R, Hillebrandt A (2008) Early Jurassic Bivalvia of northern Chile. Part III. Order Trigonioida. Beringeria 39:51–102

Philippi R (1899) Los Fósiles Secundarios de Chile. FA Brockhaus 104 pp, 42 lám. Santiago

Piatnitzky A (1933) Rético y Liásico en los valles de los ríos Genua y Tecka y sedimentos continentales de la Sierra de San Bernardo. Bol Inf Petrol 10:151–182

Piatnitzky A (1936) Estudio geológico de la región de los Ríos Chubut y Genua. Bol Inf Petrol 13:83–118

Quinzio LA (1987) Stratigraphische Untersuchungen im Unterjura des Südteils der Provinz Antofagasta in Nord. Chile. Berliner Geowiss Abh A87:1–15

Rex MA, Stuart CT, Hessler RR, Allen JA, Sanders HL, Wilson GDF (1993) Global-scale latitudinal patterns of species diversity in the deep-sea benthos. Nature 365:636–639

Rex MA, Stuart CT, Coyne G (2000) Latitudinal gradients of species richness in the deep-sea benthos of North Atlantic. Proc Nat Acad Sci U S A 97:4082–4085

Riccardi AC (2008a) The marine Jurassic of Argentina: a biostratigraphic framework. Episodes 31:326–335

Riccardi AC (2008b) El Jurásico de Argentina y sus amonites. Rev Asoc Geol Argent 63:625–643

Riccardi AC, Damborenea SE, Manceñido MO, Ballent SC (1988) Hettangiano y Sinemuriano marinos en Argentina. Actas 5° Congr Geol Chileno 2:C359–C373

Riccardi AC, Damborenea SE, Manceñido MO, Ballent SC (1991) Hettangian and Sinemurian (Lower Jurassic) biostratigraphy of Argentina. J South Am Earth Sci 4:159–170

Riccardi AC, Gulisano CA, Mojica J, Palacios O, Schubert C (1992) Western South America. In: Westermann GEG (ed) The Jurassic of the Circum Pacific. Cambridge University Press, London

Riccardi AC, Damborenea SE, Manceñido MO, Leanza HA (2011) Megainvertebrados del Jurásico y su importancia geobiológica. In: Leanza HA, Arregui C, Carbone O, Danieli JC, Vallés JM (eds) Geología y Recursos Naturales de la Provincia del Neuquén, Relatorio del 18° Congr Geol Argent, pp 441–464

Rigal R (1930) El Liásico en la Cordillera del Espinacito (Provincia de San Juan). Publ Serv Geol Nac (Argent) 74:5–9

Robbiano JA (1971) Contribución al conocimiento estratigráfico de la sierra del Cerro Negro, Pampa de Agnia, Provincia de Chubut, República Argentina. Rev Asoc Geol Argent 26:41–56

Rosen BR (1992) Empiricism and the biogeographical black box: concepts and methods in marine palaeobiogeography. Palaeogeogr Palaeoclimatol Palaeoecol 92:171–205

Roy K, Jablonski D, Valentine JW, Rosenberg G (1998) Marine latitudinal diversity gradients: tests of causal hypothesis. Proc Nat Acad Sci U S A 95:3699–3702

Roy K, Jablonski D, Martien KK (2000) Invariant size-frequency distributions along a latitudinal gradient in marine bivalves. Proc Nat Acad Sci U S A 97:13150–13155

Sanders HL (1968) Marine benthic diversity: a comparative study. Am Natural 102:243–282

Sepúlveda P, Naranjo JA (1982) Geología de la Hoja Carrera Pinto. Escala 1:100.000. Serv Nac Geol Min, Carta Geol Chile 53:1–62

Stehli FG (1968) Taxonomic diversity gradients in pole location: the recent model. In: Drake ET (ed) Evolution and Environment. Yale University Press, New Haven

Stehli FG, McAlester AL, Helsley CE (1967) Taxonomic diversity in recent bivalves and some implications for geology. Geol Soc Am Bull 78:455–466

Stehli FG, Douglas RG, Newell ND (1969) Generation and maintenance of gradients in taxonomic diversity. Science 164:947–949

Steinmann G (1881) Zur Kenntnis der Jura- und Kreideformation von Caracoles (Bolivia). N Jb Min Geol Pal BB 1:239–301

Thiele Cartagena R (1964) Reconocimiento geológico de la Alta Cordillera de Elqui. Ins Geol Fac Cienc Fis Matem Univ Chile Publ 27:133–197

Thomas H (1958) Geologia de la Cordillera de la Costa entre el valle de La Ligua y la Cuesta de Barriga. Ins Inv Geol Bol 2:1–86

Valdovinos C, Navarrete SA, Marquet PA (2003) Mollusk species diversity in the Southern Pacific: why are there more species towards the pole? Ecography 26:139–144

Valentine JW, Jablonski D (2010) Origins of marine patterns of biodiversity: some correlates and applications. Palaeontology 53:1203–1210

Vicente JC (2005) Dynamic paleogeography of the Jurassic Andean Basin: pattern of transgression and localisation of main straits through the magmatic arc. Rev Asoc Geol Argent 60:221–250

Vizán H (1988) Estudios paleontológicos y paleomagnéticos preliminares de la Formación Lepá (rio Gualjaina, Chubut). Rev Asoc Geol Argent 43:327–337

Volkheimer W, Manceñido MO, Damborenea SE (1978) Zur Biostratigraphie des Lias in der Hochkordillere von San Juan. Argentinien. Münst Forsch Geol Paläontol 44/45:205–235

Wahnish E (1942) Observaciones geológicas en el Oeste del Chubut. Estratigrafía y fauna del Liásico en los alrededores del rio Genua. Bol Serv Geol Nac 51:1–73

Waller TR (1971) The glass scallop *Propeamussium*, a living relict of the past. Am Malacol Union Ann Rep 1970:5–7

Waller TR (2006) Phylogeny of families in the Pectinoidea (Mollusca: Bivalvia): importance of the fossil record. Zool J Linnean Soc 148:313–342

Waller TR (2011) Neogene Paleontology of the Northern Dominican Republic. 24. Propeamussiidae and Pectinidae (Mollusca: Bivalvia: Pectinoidea) of the Cibao Valley. Bull Am Paleontol 381:1–195

Weaver C (1931) Paleontology of the Jurassic and Cretaceous of West Central Argentina. Mem Univ Washington 1:1–469

Chapter 5
Hemispheric Scale

Abstract On the basis of the distribution of benthonic Jurassic bivalve genera in the Southern Hemisphere paleobiogeographic units (biochoremas) were characterized according to their biologic contents (mainly levels of endemism and BSN analysis). Two first-order paleobiogeographic units are recognized for this region: Tethyan and South Pacific. Their evolution through time is followed from the Triassic to the earliest Cretaceous. The Tethyan unit was undoubtedly the most mature and persistent, with five subordinate units in this part of the world at different times: an Australian unit restricted to the Late Triassic, a North Andean unit, which appeared sporadically as an endemic center, a South Tethyan unit (Iran and Arabia), an SE Tethyan unit (Himalayas) already present in the Triassic but well established from Middle Jurassic, and an East African unit which is recognizable from Bajocian times onwards. From Late Triassic times a South Pacific first-order unit is also evident, with a persistent Maorian biochorema and a South Andean unit identifiable through most of the Jurassic. Being a transitional biogeographic setting between Tethyan and South Pacific first-order units, the South Andean one is included in the South Pacific due to the common presence of antitropical genera. The East African unit is included within the Tethyan during the Jurassic, but later, in Early Cretaceous times, it split into two units, one of which was regarded as part of the "South Temperate Realm" by Kauffman. The rank of all these units changed with time. The evolution of the paleobiogeographic patterns with time also reveals the nearly complete disruption as a consequence of the Triassic/Jurassic biotic crisis, with almost no units recognizable for the earliest Jurassic.

In this chapter, we will discuss the distribution in time and space of marine Triassic and Jurassic bivalves from the Paleo-Southern Hemisphere, using the available data for the recognition, description, and study of the evolution of paleobiogeographic units or biochoremas.

The term biochorema (biochore in the sense of (Makridin 1973; Westermann 2000a and Cecca and Westermann 2003) applies to a biogeographic unit of any rank.

S. E. Damborenea et al., *Southern Hemisphere Palaeobiogeography of Triassic-Jurassic Marine Bivalves*, SpringerBriefs Seaways and Landbridges: Southern Hemisphere Biogeographic Connections Through Time, DOI: 10.1007/978-94-007-5098-2_5, © The Author(s) 2013

Biochoremas are highly dynamic units; they may appear and disappear, expand or shrink in their geographic range, and also change in rank through time. Those units based on benthonic faunas do not necessarily have a coincident history with those based on pelagic groups. All these aspects are clearly exemplified by the history and evolution of Southern Hemisphere biogeographic units during Triassic and Jurassic times (Damborenea 2002b).

As already pointed out in Sect. 3.1, the traditional knowledge of global bivalve distribution during the Mesozoic was based mainly on data from the Northern Hemisphere. Global studies on the distribution of Mesozoic bivalves used to be largely based on databases which had a very poor coverage of the Southern Hemisphere, but this has changed now, and a wealth of recently published information from this region certainly added substantial evidence to be analyzed (see references in Damborenea 2002b).

To this historic bias, another important factor should be added: mainly as a consequence of the unevenness of land masses distribution on the Earth, the northern–southern asymmetry evident in some aspects of marine faunal distribution patterns is well known (see discussion in Crame 1996a, 2000a, b). Nevertheless, three major biogeographic units based on benthonic invertebrates (one low-latitude and two high-latitude) are recognized for Permian (see Shi and Grunt 2000 and references therein), Triassic (e.g., Diener 1916; Stevens 1980), Cretaceous (e.g., Fleming 1963; Kauffman 1973; Sohl 1987; Stevens 1980), and Cenozoic times (e.g. Fleming 1963; Hayami 1989).

In contrast, the just mentioned hemispheric asymmetry of biogeographic units during most of the Mesozoic led many authors to recognize only two first-order paleobiogeographic units for the Jurassic, mainly on the basis of the distribution of ammonites: the Boreal and Tethyan Realms (Hallam 1969, 1971, 1977; Stevens 1980, 1990; Doyle 1987; Challinor et al. 1992; Hillebrandt et al. 1992; and many others, see discussion and references in Westermann 2000a, b). As Kauffmann (1973) aptly expressed, "poor knowledge of south temperate [bivalve] faunas has led many authors to assume no 'anti-Boreal' realm existed south of Tethys during the Mesozoic" and he demonstrated that this is clearly not true for the Cretaceous, although he admitted that "this was possibly true in the Early Jurassic". Furthermore, Ager (1975, p. 17) said that "it is often commented that there was no southern counterpart" of the Boreal Realm in the Jurassic, but certain invertebrates do seem to be "restricted to that region". In fact, lack of data and proper analysis for the Southern Hemisphere pervaded also studies of other fossil groups (see for instance Dommergues et al. 2001), and was explicitly acknowledged by Crame (1986) in relation to bivalves.

This assumed twofold division, most probably heavily influenced by leading opinions based on ammonite distribution (for instance Arkell 1956) hindered the determination of the possible role that the austral regions may have had in the origin and diversification of the biota (see Crame 1997) during the Jurassic. Widespread phenomena such as austral endemism or antitropicality were either ignored or their relative importance was disregarded. Some authors who work on Southern Hemisphere faunas (especially bivalves) have been challenging this twofold division for a long time (see Stevens 1980 and references therein; Crame 1986, 1987, 1992, 1993, 1993; Damborenea 1993, 1996, 2002a, b; Enay and Cariou 1997 and references therein). As a result, several marine paleobiogeographic units of low rank were proposed for the Southern Hemisphere Jurassic on the basis of the known distribution of different marine organisms. The relationships, rank, and history of these units were reviewed by Enay and Cariou (1997) and Westermann (2000b).

Jurassic paleobiogeographic provinces based on bivalves were analyzed quantitatively for the European Tethyan and Proto-Atlantic (Liu 1995; Liu et al. 1998). For the Southern Hemisphere, several contributions related to paleobiogeographic issues using these organisms were already available for the South Pacific (Stevens 1967, 1977, 1980, 1989, 1990; Hayami 1984, 1987; Grant-Mackie et al. 2000), Antarctica (Crame 1987, 1992, 1996a, b), and the South American margin of the Pacific (Damborenea and Manceñido 1979, 1988; Hillebrandt 1981; Hallam 1983; Damborenea 1993, 1996). A first comprehensive paleobiogeographic analysis based on late Triassic–Jurassic bivalves for the Southern Hemisphere was provided by Damborenea (2002b). We discuss here the issue of paleobiogeographic units based on the distribution of bivalves in the light of new data and the application of new techniques.

5.1 Data

Occurrences of Triassic and Jurassic bivalve species were compiled from various published sources as well as the author's own data, and plotted stage by stage from Induan to Berriasian.

Data were then gathered within wide areas, each containing a large variety of habitats. About 14 such areas were chosen for the present analysis (located in Fig. 5.1). On the whole, they represent a wide coverage of the Southern Hemisphere Triassic and Jurassic seas, but some important gaps still exist, limited by the availability of data. Both the amount and quality of data are, unsurprisingly, very uneven, but although this hinders serious detailed quantitative analysis, the database provides enough information to obtain a broad framework.

For this analysis data were processed at the genus group level and according to the following nine time intervals: Induan-Anisian, Ladinian-Carnian, Norian-Rhaetian, Hettangian-Sinemurian, Pliensbachian-Toarcian, Aalenian-Bajocian, Bathonian-Callovian, Oxfordian-Kimmeridgian, and Tithonian-Berriasian (Table 5.1). Admittedly, this implies a loss of detail in the information for some regions, but on the other hand it allows the use of some occurrences with uncertain stratigraphic provenance. The age slice including Tithonian and Berriasian has the extra drawback that in this way any event related to the Jurassic-Cretaceous boundary may pass unnoticed. Data are summarized in Table 5.2.

Endemic genera were recognized as such for each time interval. Since geographic ranges of taxa may change through time, sometimes this results in a different categorization for the same taxon. For instance, a genus may be endemic

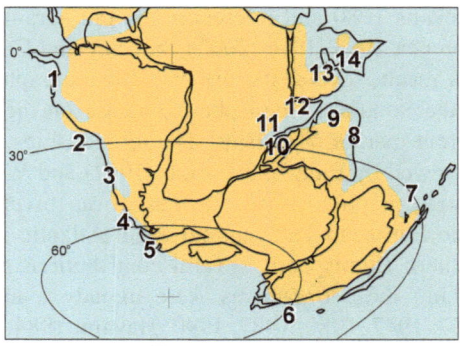

Fig. 5.1 Location of the geographic units used in this study. *1* NW South America (Colombia), *2* Perú and northernmost Chile, *3* Central Argentina and Chile, *4* Southern Argentina and Chile, *5* Antarctica, *6* New Zealand-New Caledonia, *7* Western Australia-New Guinea, *8* Himalayas (N India, S Tibet), *9* Western India, *10* Madagascar, *11* SE Africa (Kenya, Tanzania), *12* E Africa (Eritrea, Ethiopia, Somalia), *13* Arabian Peninsula (Saudi Arabia, Yemen, Oman); and *14* Iran. Base map as in Fig. 1.2

to a certain region during a stage and then become widespread. An example is *Kalentera*, which was endemic to the Maorian Province in Late Triassic times but became antitropical during the Early Jurassic. Similarly, *Gryphaea* was restricted to high latitudes during the Triassic but later became pandemic. Conversely, a previously widespread taxon may have a temporarily restricted distribution. The recognition of these subtle changes is highly dependent on a very detailed knowledge of recorded distributions, and is thus not always easy to establish in the fossil record. Nevertheless, these changes were herein recognized and taken into account as far as possible.

A few comments about each region and the data sources are given below; numbers refer to those located in Fig. 5.1 and used throughout in the text.

5.1.1 South America

1. NW South America: (Colombia and northernmost Perú): Though relatively diverse faunas are known to occur, the published information from this region is very scarce and is not comparable with the other data sets, being confined to a few records from Norian to Toarcian and Tithonian-Berriasian. The time location of some of these data is still controversial, as clearly exemplified by the fact that the Batá Formation, which contains bivalves originally referred to the Jurassic by Bürgl (1961) and Geyer (1973), is now regarded as Cretaceous in age (Etayo Serna et al. 2003). Data used here were critically compiled from Geyer (1973, 1979) and Guzmán (1984) and references therein.

Table 5.1 Time frame used in this study for the discussion of paleobiogeographic units and their evolution, right hand column indicates the letter symbols used throughout the text to indicate the time slices used in the analysis

SYSTEM/ PERIOD	SERIES/ EPOCH	STAGE	Time slices
CRETACEOUS ↑		Valanginian	
		Berriasian	T
JURASSIC	LATE	Tithonian	
		Kimmeridgian	O
		Oxfordian	
	MIDDLE	Callovian	B
		Bathonian	
		Bajocian	A
		Aalenian	
	EARLY	Toarcian	P
		Pliensbachian	
		Sinemurian	H
		Hettangian	
TRIASSIC	LATE	Rhaetian	N
		Norian	
		Carnian	L
	MIDDLE	Ladinian	
		Anisian	I
	EARLY	Olenekian	
		Induan	
PERMIAN ↓		Changhsingian	

2. Perú and Northern Chile: Comprises most of Perú and the northernmost regions of Chile (up to about 26° S present-day latitude). References are few and not updated, and there is evidence to suggest that they do not accurately reflect the actual bivalve diversity in the area. Data sources for the Early Jurassic of Chile listed in Damborenea 1996, with the addition of data from other ages and from Perú (and some new references) in: Jaworski 1915, 1922; Körner 1937; Cox 1949, 1956; Harrington 1961; Pérez and Reyes 1977, 1983, 1985, 1986, 1991; Hayami et al. 1977; Westermann et al. 1980; Chong and Hillebrandt 1985; Prinz 1985; Riccardi et al. 1990a, b; Pérez et al. 1987; Romero et al. 1995; Aberhan and Hillebrandt 1996, 1999; Rubilar 1998; Aberhan 2007.

3. Central Argentina and Chile: This region extends from 26° to 41° S present-day latitude, and the southern part includes what is regionally known as the Neuquén basin (s.l.). The geology of this area is well known and paleogeographic maps at different moments of the time interval are available. Bivalve faunas are very well documented for the whole Jurassic, but Triassic ones are not so well known. Early Jurassic references listed in Damborenea 1996, with the addition of: Burckhardt 1900a, b, 1903; Haupt 1907; Jaworski 1914, 1915; Stehn 1923; Fuenzalida Villegas 1937; Leanza 1941; Lambert 1944;

Table 5.2 Chronologic and geographic distribution of Triassic and Jurassic bivalve genus group taxa in the areas and age intervals considered in this study

a		Regions													
		1	2	3	4	5	6	7	8	9	10	11	12	13	14
Time	I	–	2	–	–	–	4	3	12	5	–	–	–	4	3
	L	–	3	–	–	–	26	3	9	1	–	–	–	–	1
	N	13	50	19	–	–	37	26	49	1	–	–	–	21	69
	H	15	63	76	–	5	13	–	17	–	–	–	–	–	–
	P	3	75	102	53	2	25	5	10	–	15	18	–	4	25
	A	–	33	69	–	20	51	24	12	10	13	30	–	23	10
	B	–	1	31	–	23	38	1	23	130	50	49	57	66	21
	O	–	5	11	–	34	30	7	13	32	16	76	49	45	–
	T	15	3	70	27	25	19	4	23	15	3	29	–	9	–

b		Regions													
		1	2	3	4	5	6	7	8	9	10	11	12	13	14
Time	I	–	2	–	–	–	1	3	8	3	–	–	–	4	3
	L	–	2	–	–	–	14	1	6	1	–	–	–	–	0
	N	8	32	15	–	–	18	11	20	0	–	–	–	10	33
	H	12	37	41	–	4	10	–	14	–	–	–	–	–	–
	P	2	42	54	30	2	16	3	8	–	11	19	–	1	17
	A	–	22	44	–	14	31	19	12	3	9	20	–	16	8
	B	–	1	22	–	13	21	1	18	57	31	30	37	38	11
	O	–	4	9	–	19	20	3	6	18	10	47	32	26	–
	T	8	0	39	16	15	10	2	15	8	2	17	–	8	–

c		Regions													
		1	2	3	4	5	6	7	8	9	10	11	12	13	14
Time	I	–	0	–	–	–	2	0	1	0	–	–	–	0	0
	L	–	0	–	–	–	6	0	0	0	–	–	–	–	0
	N	1	3	0	–	–	8	4	4	0	–	–	–	0	8
	H	0	3	3	–	0	0	–	1	–	–	–	–	–	–
	P	0	6	5	1	0	2	0	0	–	0	0	–	0	0
	A	–	2	4	–	0	7	0	0	1	0	0	–	1	0
	B	–	0	2	–	1	5	0	1	8	1	2	1	3	1
	O	–	1	2	–	1	2	0	2	3	2	5	1	0	–
	T	0	1	5	1	1	1	0	3	1	0	3	–	0	–

a Total number of genera, b Number of cosmopolitan (s.s.) genera, c Number of endemic genera. *Regions*: key to numbers in Fig. 5.1; *time slices*: key to letters in Table 5.1

Sokolov 1946; Levy 1967; Thiele Cartagena 1967; Cecioni and Westermann 1968; Hallam et al. 1986; Lo Forte 1988; Damborenea 1990, 1998, 2002a, 2004; Leanza and Garate 1987; Riccardi 1988; Riccardi et al. 1990a, b, c, 1997, 2004, 2011; Damborenea et al. 1992; Leanza 1993; Aberhan 1994, 2004, 2007; Malchus and Aberhan 1998; Rubilar 1998; Damborenea and Manceñido 2005, 2012; Damborenea and Lanés 2007; Pérez et al. 2008; and unpublished data.

4. Southern Argentina and Chile: Includes early Jurassic (Pliensbachian-Toarcian) deposits in Chubut Province (Argentina), which are considered by some authors as a southern extension of the Neuquén basin, and late Jurassic-early Cretaceous (Tithonian-Berriasian) beds from the Austral Basin of southernmost Chile and Argentina. Very few data are known, references for Chubut listed in Damborenea 1996, with the addition of: Feruglio 1936; Leanza 1968; Riccardi 1977; Olivero 1988; and unpublished data.

5.1.2 Antarctica, New Zealand-New Caledonia

5. Antarctica: This region comprises mainly the Antarctic Peninsula and adjacent areas. Data from the Falkland Plateau have also been included here. Knowledge is abundant but very patchy and scattered, taking into account the large size of the area. Data were compiled from (Stevens 1967; Thomson and Willey 1972; Thomson 1975a, b, 1981, 1982; Willey 1975a, b, c; Jones and Plafker 1976; Quilty 1978, 1982, 1983; Edwards 1980; Crame 1981, 1982a, b, 1983, 1984, 1985, 1996c; Jeletzky 1983; Medina and Ramos 1983; Thomson and Tranter 1986; Riccardi et al. 1990c; Doyle et al. 1990; Thomson and Damborenea 1993; Crame et al. 1993; Crame and Kelly 1995; Kelly 1995; Riley et al. 1997). There is a lot of information which remains unpublished in Hikuroa's thesis (2005).
6. New Zealand-New Caledonia: Although these regions are now widely separated in space, in the past they were closely related and their Mesozoic bivalve fauna was very alike and studied by the same authors. New Zealand faunas are by far better known, only few data are available from New Caledonia, and thus data from these two areas are pooled together for the analysis. Bivalve diversity is very well documented for the whole time interval of the present study. Some bivalves are often used in biostratigraphy and they define several units. Buchiid species, for instance, are used to subdivide and correlate within the Puaroan Stage (Hikuroa and Grant-Mackie 2008 and references therein). Data were compiled from Trechmann 1918, 1923; Marwick 1935, 1953, 1956; Avias 1953; Fleming 1959, 1962, 1964, 1987; Grant-Mackie 1960, 1976a, b, 1978a, b, c, d, 1980a, b, 2011; Speden 1970; Stevens 1978; Freneix et al. 1974; Speden and Keyes 1981; Begg and Campbell 1985; Crampton 1988; Grant-Mackie and Silberling 1990; Damborenea and Manceñido 1992; Damborenea 1993; Gardner and Campbell 1997, 2002, 2007; Gardner 2005, 2009; Hikuroa and Grant-Mackie 2008; and unpublished data (in Hudson 1999).

5.1.3 Australia-New-Guinea

7. Western Australia (Bonaparte, Carnarvon, and Perth Basins), New Guinea, and
   Sula Islands: Although there are a few scattered data from most of the time
   intervals considered, only those for Norian-Rhaetian and Aalenian-Bajocian
   faunas are used here. Data were compiled from: Etheridge 1910; Whitehouse
   1924; Teichert 1940; Skwarko 1967, 1973, 1974, 1981a, b, 1983; Coleman and
   Skwarko 1967; Skwarko et al. 1976; Sato et al. 1978; Grant-Mackie 1994;
   Waterhouse 2008.

5.1.4 Southern Asia

8. Himalayas (northern India, Afghanistan and southern Tibet): Rossi-Ronchetti
   and Fantini-Sestini 1961; Rossi-Ronchetti 1970; Li and Grant-Mackie 1988,
   1994 and references therein; Li 1990; Yancey et al. 2005.
9. Western India: No data for late Triassic and Early Jurassic are available. On the
   other hand, knowledge of middle Jurassic bivalve faunas is very good and has
   been the subject of several recent monographs. Data were compiled from
   Kitchin 1903; Cox 1935b, 1940, 1952; Agrawal 1956a, b; Kanjilal and Singh
   1973; 1980; Singh and Kanjilal 1974, 1977, 1982; Agrawal and Rai 1978;
   Kanjilal 1979a, b, 1980a, b, 1981; Singh and Rai 1980; Singh et al. 1982; Jaitly
   and Singh 1983; Pandey and Agrawal 1984; Jaitly 1986a, b, c, 1988, 1989,
   1992; Jaitly et al. 1995; Pandey et al. 1996; Fürsich and Heinze 1998;
   Fürsich et al. 2000.

5.1.5 Africa

10. Madagascar: Reliable data are available only from Toarcian times onwards.
    Due to the scarcity of data, some records that appear to be trustworthy were
    included despite the fact that they are not yet backed by descriptions and
    figures (for instance in Mette 2004). See Geiger and Schweigert (2006) for a
    revision of Jurassic sedimentary cycles and tectonic history in western
    Madagascar. Data compiled from Newton 1889, 1895; Douvillé 1904;
    Thevenin 1908; Barrabé 1929; Besairie 1930; Besaire and Collignon 1972;
    Nicolaï 1950–1951; Mette 2004.
11. Southeast Africa (Kenya and Tanzania): Again, there are no data older than
    Toarcian, and were compiled from Weir 1930; Dietrich 1933; Cox 1965;
    Aberhan et al. 2002; Bussert et al. 2009.
12. East Africa (Ethiopia, Somalia, and Eritrea): Reliable data only from
    Pliensbachian times onwards, but some of the units are not precisely dated yet;
    see for instance the discussion for the Tendaguru Formation units in Bussert

et al. (2009). Data for the analysis were compiled from Fütterer 1897; Dacque 1905; Basse 1930; Díaz-Romero 1931; Cox 1935a and references therein; Stefanini 1939; Venzo 1949; Jaboli 1959; Ficcarelli 1968; Jordan 1971; Abbate et al. 1974; Kiessling et al. 2011.

5.1.6 Near East

13. Southern Arabian Peninsula: Yemen, Oman, Saudi Arabia: Published data are very uneven for southern Arabic peninsula. Bivalves from this region have general affinities with those from eastern Africa, but they are in need of revision. The geology is well known in relation to oil industry; see for instance (Ziegler 2001) for paleogeographic synthesis and tectonic history. Data were compiled from (Basse 1930; Arkell 1956; Hudson and Jefferies 1961; El-Asa'ad 1989; Manivit et al. 1990; Howarth and Morris 1998; Krystyn et al. 2003; Yancey et al. 2005).
14. Iran: Bivalve faunas from central Iran beds are well known, especially those from the late Triassic. The territory was in the Southern Hemisphere during that time, but it then migrated northwards to the paleoequator. Bivalve data were compiled from (Fischer 1915; Cox 1936; Fantini-Sestini 1966; Geyer 1977; Kluyver et al. 1978; Kristan-Tollmann et al. 1980; Kalantari 1981; Schairer et al. 2000; Hautmann 2001a, b; Fürsich et al. 2005).

5.2 Biochorema Recognition

5.2.1 Biogeographic Units and Their Characterization

The scarcity of data for the Early and Middle Triassic makes it impossible to attempt a paleobiogeographic analysis based on endemism. Most of the regions had less than 10 genera for each time interval from Induan to Carnian (see Table 5.2), and were not included in the analysis, which was performed for Norian times onwards.

Although always low, at different times through the Late Triassic and Jurassic, endemism within the different areas of the Southern Hemisphere varied, and was used here to recognize and characterize paleobiogeographic units. For the following analysis, percentage of endemism was calculated excluding cosmopolitan forms (as done by Kauffman 1973) and including only strictly endemic taxa. In the following discussion, all percentages quoted are calculated over total minus cosmopolitan genera.

A maximum of seven basic biochoremas (not all of them persistent during the whole time range) and two units of higher rank were recognized for the different time slices considered according to the distribution and percentage of endemic taxa (Table 5.3) (Damborenea 2002b, modified here). Some of the basic units are regarded as belonging to the Tethyan first-order biochorema, while others can be grouped in another high rank unit which following Westermann's (2000b) recommendations is called South Pacific (Challinor 1991, Austral Realm in Damborenea 1993). Overall endemism for second-order biochoremas of this last high rank unit is proportionally high during the latest Triassic, between 8 and 22 % for the Early Jurassic, and over 14 % for Middle and Late Jurassic. These fluctuations suggest its change in rank through time, as will be discussed later.

These percentages are low according to normalized scales based on species distribution (see discussion in Westermann 2000a), but are of the same order as those used by other authors to define paleobiogeographic units on the basis of fossil bivalve genera (e.g., Kauffman 1973), and they allow the recognition of units of different rank. Nevertheless, taking into account the uneven nature of the database, no attempt is made here to establish a threshold of minimum values for each rank. It is interesting to remember that the distinction between Tethyan and Boreal Realms based on bivalves in the Northern Hemisphere rests on only a few taxa. According to Liu (1995), for instance, for the Pliensbachian only the presence of *Hippopodium* and *Meleagrinella* is characteristic of the Boreal Realm and *Weyla* and *Lithiotis* of the Tethyan Realm in Europe.

All units are further characterized by other aspects, such as the presence and relative abundance of taxa with high-latitude or strictly low-latitude affinities (Table 5.3), overall diversity, and the presence/absence of certain higher rank taxonomic groups. Antitropical taxa were common within monotoids, pectinoids, inoceramoids, and other bivalve groups, and add character to some of these units.

On a worldwide scale, this present arrangement implies the presence of three first-order units during the Jurassic based on bivalve data: Boreal, Tethyan, and South Pacific (= Austral). The distribution of bivalves and the corresponding proposed paleobiogeographic zonation is evidently not symmetric relative to the paleoequator. In fact, in view of the unbalanced distribution of land/water masses and the related uneven oceanic current patterns (which are mostly hypothetic at this stage), such a paleobiogeographic asymmetry is only to be expected. Nevertheless, it is quite clear that the analyzed data do not support the alternative of overstressing this asymmetry to the point of reducing the paleobiogeographic zonation for the Jurassic to only two first-order units (Damborenea 2002b).

At the family level, all three first-order paleobiogeographic units are well characterized. During the Triassic, Jurassic, and Early Cretaceous most genera of Anomiidae, Burmesiidae, Ceratomyopsidae, Cuspidariidae, Diceratidae, Dicerocardiidae, Isoarcidae, Lithiotidae, Mactromyidae, Malleidae, Megalodontidae, Myalinidae, Myopholadidae, Myophoricardiidae, Mysidiellidae, Ostreidae, Pergamidiidae, Protocardiidae, Ptychomyidae, Pulvinitidae, Requieniidae, Sowerbyidae, and Unicardiopsidae were restricted to low latitudes and characterized the Tethyan Realm. To these, a group of families with more than average low-latitude taxa (Fig. 5.2) should be added: Laternulidae, Pinnidae, Prospondylidae, Tancrediidae, Arcticidae, and Myophoriidae. The first four of these have no high-latitude taxa, and the others contain less than average high-latitude genera.

Table 5.3 Southern Hemisphere second-order paleobiogeographic units recognized and their percentage of endemic generic level taxa through time (bottom row)

Time	Biochoremas		Percentages (%)			
	First order	Second order	Low-latitude	High-latitude	Trans-temperate	Endemics
Norian–Rhaetian	South Pacific	Maorian [6]	5	37	16	42
		Andean transitional [3]	40	40	20	0
	?	North Andean [1 + 2]	60	15	10	15
	Tethyan	S Tethyan [13 + 14]	80	0	0	20
		SE Tethyan [8 + 9]	86	0	0	14
		Australian [7]	73	0	0	27
Hett–Sinem	South Pacific	Maorian [5–6]	25	25	50	0
		South Andean [2 + 3]	59	19	14	8
	Tethyan	North Andean [1]	33	0	67	0
		SE Tethyan [8]	67	0	0	33
Pliens–Toarc	South Pacific	Maorian [5 + 6]	0	45	33	22
		South Andean [4]	48	22	26	4
	Tethyan	North Andean [1 + 2 + 3]	54	18	15	13
		S Tethyan [13 + 14]	80	0	20	0
		SE Tethyan [7 + 8]	75	0	25	0
		E African [10 + 11]	89	0	11	0
Aalenian–Bajocian	South Pacific	Maorian [5 + 6]	40	24	8	28
		South Andean [3]	56	20	8	16
	Tethyan	North Andean [2]	73	9	0	18
		S Tethyan [13 + 14]	87	0	0	13
		SE Tethyan [7 + 8 + 9]	84	0	8	8
		East African [10 + 11]	100	0	0	0
Bath-Call	South Pacific	Maorian [5 + 6]	61	5	5	29
		South Andean [3]	45	33	0	22
	Tehyan	East African [10 + 11 + 12 + 13]	82	2	6	10
		SE Tethyan [8 + 9]	82	1	5	12

(continued)

Table 5.3 (continued)

Time	Biochoremas		Percentages (%)			
	First order	Second order	Low-latitude	High-latitude	Trans-temperate	Endemics
Oxfor-Kimm	South Pacific	Maorian [5 + 6]	29	41	12	18
		South Andean [3]	0	0	0	?
	Tehyan	East African [10 + 11 + 12 + 13]	74	9	4	13
		SE Tethyan [8 + 9]	52	19	5	24
Tithon-Berr	South Pacific	Maorian [5 + 6]	29	36	21	14
		South Andean [3 + 4]	67	8	11	14
	Tehyan	East African [10 + 11 + 13]	72	0	7	21
		North Andean [1]	100	0	0	0
		SE Tethyan [8 + 9]	59	7	7	27

The percentage of low-latitude, high-latitude, and trans-temperate genera is also indicated. All percentages are calculated over the number of total minus cosmopolitan genera. Numbers in brackets behind the name of each biochorema refer to the geographic units used here, see location in Fig. 5.1

Fig. 5.2 Percentage of genera with different paleobiogeographic distribution during the Triassic and Jurassic within selected bivalve families. Families arranged according to the percentage of high-latitude genera. Average composition for the whole time interval at the *top*, background diagonal lines indicate average percentage of low-latitude (*orange*) and high-latitude (*red*) genera. Own data based on nearly 500 bivalve genera

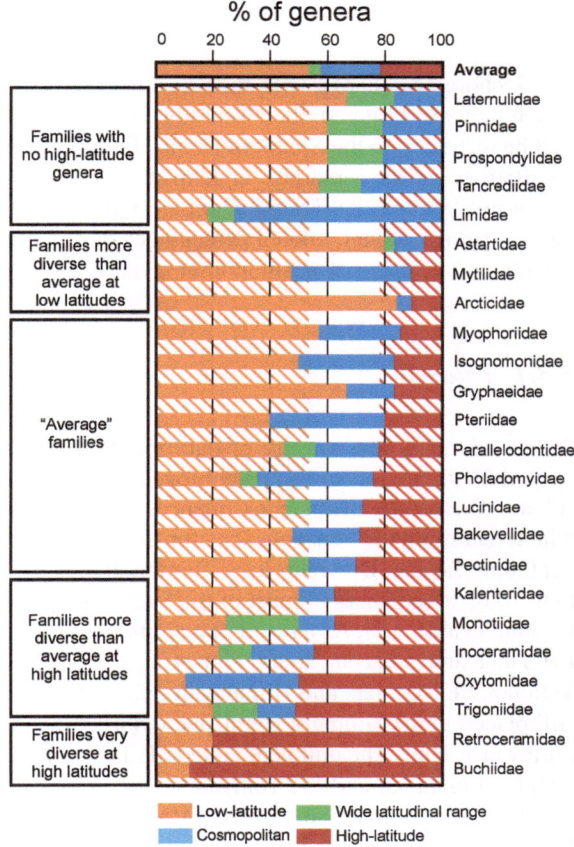

Families strictly restricted to high latitudes in the Jurassic are fewer; among them Asoellidae, Minetrigoniidae, and Palaeopharidae are present in both Boreal and South Pacific Realms whereas Sportellidae and Yoldiidae are known only from the Boreal Realm. There is a consistent group of families which have more than average high-latitude taxa and at the same time less than average low-latitude genera (Fig. 5.2), all of them diverse and abundant in both Boreal and South Pacific Realms: Kalenteridae, Monotidae, Inoceramidae, Oxytomidae, Trigoniidae, Retroceramidae, and Buchiidae. It should be pointed out here that updated systematic knowledge of bivalve faunas is still very uneven in the Southern Hemisphere and that future revisions are likely to alter this picture slightly.

5.3 Evolution of Biochoremas

As biochoremas change with time in rank, geographic spread, and even realm to which they are assigned, the discussion below will follow a stratigraphic order and will be restricted to the Southern Hemisphere. The discussion will include different qualitative aspects, mainly endemism, as well as the quantitative analysis performed around the Triassic-Jurassic boundary (cluster analysis and BSN). In the following account, quoted percentages of different elements of the faunas were calculated exclusive of cosmopolitan genera, as done by Kauffman (1973).

5.3.1 Triassic

As already said, the scarcity of data for the Early and Middle Triassic makes it very difficult to recognize paleobiogeographic patterns in the Southern Hemisphere. This is one of the consequences of the vast end-Permian extinction event, which had dramatic consequences for the systematic and ecologic composition of benthonic faunas. Diversity loss after the extinction resulted in a reduction to less than 10 bivalve genera in most of the Southern Hemisphere regions from Induan to Carnian (Table 5.2). As a result of a similar analysis of bivalve generic distribution for Tibet and adjacent areas, Niu et al. (2011) recognized three provinces (NE Tethyan, SE Tethyan, and Himalayan), all of them Tethyan. Only the last two were still placed in the Southern Hemisphere during the early Triassic, they all migrated northwards later. The unit here called SE Tethyan roughly corresponds to Niu et al. (2011) Himalayan and SE Tethyan.

Nonetheless, already in the early Middle Triassic two endemic genera are known from the Etalian of New Zealand (*Etalia* Begg and Campbell and *Marwickiella* Sha and Fürsich) in an otherwise low diverse fauna. These indicate the incipient appearance of an endemic center in this circum-polar area, which would develop later on. Not surprisingly, the most diverse Southern Hemisphere fauna for this time was located in low paleolatitudes, in the Himalayan region, but endemism there was restricted to only one doubtful genus. Niu et al. (2011), mostly based on relative diversity, proposed the continuity of the three provinces (NE Tethyan, SE Tethyan, and Himalayan) in the Tibetan area during the whole Triassic.

This emerging pattern became stronger by Ladinian and Carnian times, when all the endemic genera known so far for the Southern Hemisphere lived in the southernmost region (New Zealand-New Caledonia): *Praegonia* Fleming and *Agonisca* Fleming from the early Kaihikuan, later *Balantioselena* Speden and *Manticula* Waterhouse, while the first *Hokonuia* Trechmann are already Oretian. This supports the definite establishment of a Maorian biochorema by mid-Triassic times. Other faunas of that age are very poorly known. The three Tibetan provinces are still recognized by Niu et al. (2011), but by this time only the southern one (Himalayan) remains in the Southern Hemisphere.

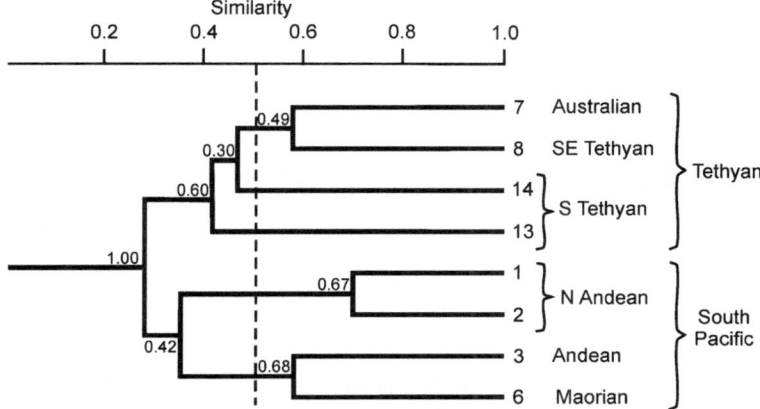

Fig. 5.3 Cluster analysis of South Hemisphere regions for the Norian-Rhaetian interval, values on each node representing the support value for the node obtained by bootstrapping; cophenetic correlation: 0.694. Similarity measure: Simpson's coefficient; algorithm: paired group, number of iterations for the bootstrapping: 1000. Regions code-numbered as in Fig. 5.1. To the right: paleobiogeographic units as recognized here

Endemism significantly increased in the whole hemisphere toward the Late Triassic, and up to six basic units are recognizable (Damborenea 2002b). The South Pacific first-order biochorema includes a very distinctive Maorian unit, to which perhaps an Andean Transitional unit could be added. All other biochoremas are included in the Tethyan Realm, with an Australian unit and two eastern Tethyan units (here called S Tethyan and SE Tethyan, this last one corresponding to the Himalayan province of Niu et al. 2011).

The affinities of the North Andean endemic center are obscure; this unit could be related to the South Pacific on the basis of the obtained dendrogram (Fig. 5.3) and BSN analysis (Fig. 5.4), but has also many links to the Tethyan unit.

The Maorian (Diener 1916) biochorema presents very high endemism: the genus *Hokonuia* persists from the previous stage and becomes abundant, *Caledogonia* Freneix and Avias and other local genera (*Heslingtonia* Fleming, *Kalentera* Marwick, *Maorimonotis* Grant-Mackie, *Oretia* Marwick, *Ouamouia* Campbell, *Torastarte* Marwick) are also endemic to New Zealand-New Caledonia during this time. This biochorema is very well characterized during the Middle and Late Triassic, not only by the percentage of endemics (42 %), but also by the abundance (37 %) of genera with antitropical or austral distribution, such as *Eomonotis* Grant-Mackie, *Inflatomonotis* Grant-Mackie, *Maoritrigonia* Fleming, *Minetrigonia* Kobayashi and Katayama, and *Triaphorus* Marwick. Strictly low-latitude genera are very few (5 %).

An Australian biochorema had very low diversity but is also well characterized by several endemic genera (27 %), among them *Gervillancea* Skwarko, *Guineana* Skwarko, and *Somareoides* Skwarko. Low-latitude genera are more numerous, but

Fig. 5.4 BSN of South
Hemisphere regions for the
Norian-Rhaetian interval.
Continuous thin lines:
distance value ≥0.25 and
<0.50 (75 % to more than
50 % of genera shared);
dashed lines: distance value
≥0.50 (50 % of genera
shared or less); support
values written on each line

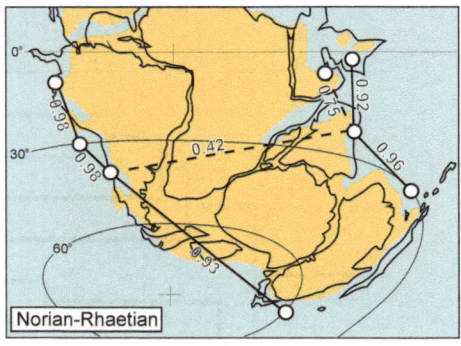

high-latitude and trans-temperate forms are absent, clearly indicating a fully
Tethyan affinity of the faunas of this region.

A North Andean endemic center extends along the Andes from Colombia to
northern Chile. Most of the bivalve fauna has low-latitude affinities, but the
endemic genera *Isopristes* Nicol and Allen and *Schizocardita* Körner are abundant
elements of the Late Triassic fauna from Peru and northern Chile (totaling 15 %),
and some genera with high-latitude distribution are also present (15 %).

The central Argentina-Chile area represents a transitional zone, with no endemic taxa but with a high percentage (40 %) of high-latitude genera, including
Maoritrigonia and *Minetrigonia*. The abundance of trans-temperate taxa (20 %) is
mainly due to genera distributed along the eastern margins of the Paleo-Pacific in
both hemispheres. Reference of both the North Andean and Andean Transitional
biochoremas to the Tethyan or the South Pacific Realm is thus still uncertain on
these evidences alone.

Apart from endemism, the quantitative analysis also shows that by the end-
Triassic there was a clear differentiation among regions. Cluster analysis (Fig. 5.3)
resulted in three minor groups with similarity coefficients higher than 0.50:
northern South America (= North Andean endemic centre), southern South
America + New Zealand, and Australia + Himalayas, although this last grouping
is not strongly supported (support value just lower than 0.50). The dendrogram
supports the idea of two major biochoremas: the first one grouping all the Pacific
localities (west coast of South America and New Zealand) while the other includes
the Tethyan localities (Iran, Arabia, Himalayas and Australia). On this evidence,
the Andean Transitional biochorema (and perhaps also the North Andean one) can
be related to the South Pacific Realm.

Bootstrapped Spanning Network (BSN) (Fig. 5.4) seems to perform better on
these data, since it clearly separates both first-order biochoremas and also reflects
the strong similarities among northern and southern South America. The best
connection between both first-order biochoremas shows a high dissimilarity value
(0.58, meaning only 42 % of genera shared between localities) and a low support
value (0.42).

Fig. 5.5 Paleobiogeographic sketch for the Norian-Rhaetian time slice, showing inferred extension of second-order biochoremas. Thick broken line: approximate position of the boundary zone between Tethyan and South Pacific first-order biochoremas. Base map as in Fig. 1.2

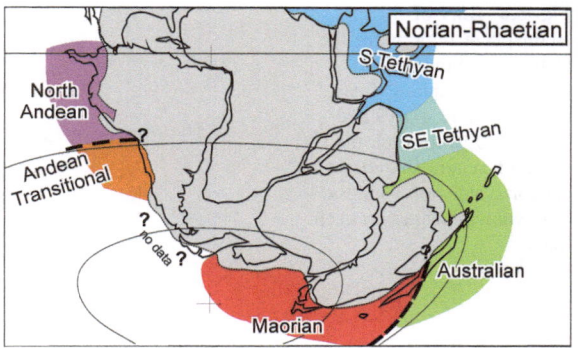

Extensive areas without late Triassic outcrops or reliable paleontologic data in the key regions prevent adequate discussion of the boundary between biochoremas at this time. Nevertheless, it is evident that paleobiogeographic patterns were mature by Norian-Rhaetian times (Fig. 5.5).

5.3.2 Early Jurassic

At the beginning of the Jurassic, the percentage of endemism was remarkably low for all regions. Although it is difficult to point a definite cause, this situation is probably a consequence of the end-Triassic extinction (see Sect. 5.5). It is interesting to note that, on the available data, the low latitude regions of the Southern Hemisphere also contain almost no endemic genera. The seemingly high value (33 %) in the Himalayas is in fact based on the apparent persistence into the Hettangian of only one endemic genus (*Persia* Repin) in an otherwise low diversity fauna with high proportion of cosmopolitan taxa. On the other hand, in transitional zones, such as the South Andean areas (2 and 3 here), an endemic center developed during Sinemurian times, with a few endemic genera (8 %), such as *Gervilletia* Damborenea, *Groeberella* Leanza and *Quadratojaworskiella* Reyes and Pérez. This endemic center is regarded here as part of the South Pacific unit despite the relatively high proportion of low-latitude taxa, due to the common occurrence of genera with antitropical or austral distributions (19 %), such as *Agerchlamys* Damborenea, *Asoella* Tokuyama, *Harpax* Parkinson, *Kalentera*, *Palaeopharus* Kittl, and *Kolymonectes* Milova and Polubotko (Damborenea 1998).

The quantitative study also suggests that the end-Triassic extinction event noticeably affected bivalve biogeography. Although a few localities had enough data to perform the analysis, the results show a clear homogenization of bivalve faunas. Cluster analysis on the Hettangian-Sinemurian data could not reveal major groupings among regions, with all nodes with similarity values higher than 0.50 and none of them with a significant support (values below 0.50). The BSN (Fig. 5.6) shows an evident decrease in dissimilarity values while all edges appear strongly supported (even some of the removed edges).

Fig. 5.6 BSN of South
Hemisphere regions for the
Hettangian-Sinemurian
interval. Continuous lines:
distance value ≥0.10 and
<0.25 (90 % to more than
75 % of genera shared);
support values of 1 unless
otherwise indicated on each
line

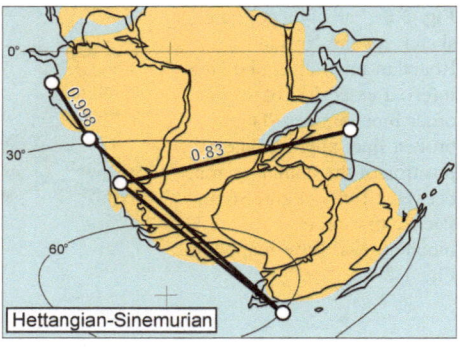

By Pliensbachian-Toarcian times the Maorian unit (= South West Pacific Province in Hallam 1977) had recovered (Fig. 5.8), having 22 % of endemic genera, including *Torastarte* and *Pseudaucella* Marwick (Grant-Mackie et al. 2000) and 44 % of high-latitude genera. A South Andean endemic center persists with a low percentage of endemics (4 %), but with a high percentage of high-latitude taxa (22 %), which include *Asoella, Harpax, Kolymonectes, Palaeopharus, Ochotochlamys* Milova and Polubotko, *Praebuchia*? Zakharov, and *Radulonectites* Hayami (Damborenea and Manceñido 1992; Damborenea 1993, 2002b). This interval is also characterized by a very high percentage of high-latitude taxa (22 %) and also the highest levels of trans-temperate (mainly east Pacific) taxa for the whole Jurassic (26 %). A North Andean endemic center is again discernible (13 % of endemic genera), with its southern limit extending farther south than earlier in the Lower Jurassic (see discussion in Sect. 4.3 and Fig. 5.11). The peculiar genera *Lithiotis* Gümbel and *Opisoma* Stoliczka, characteristic of the Jurassic tropical regions during Pliensbachian times (see discussions in Sect. 3.2.4), reached northern Chile during the Toarcian. While *Lithiotis* was absent further south in the South Andean unit, other elements of this association, such as *Opisoma* (Fig. 3.4a) and *Gervilleioperna*, extended further south (Aberhan and Hillebrandt 1999).

Cox (1965) mentioned that the presence of the species *Weyla ambongoensis* (Thevenin) in Toarcian beds of eastern Africa "affords a somewhat meagre evidence that a faunal sub-province comprising the western part of the present Indian Ocean region and extending over northern Africa had come into existence". This is a remarkable observation, taking into account the evolution of paleobiogeography of this part of Africa later in the Jurassic. Nevertheless, data from this area are scarce, and no endemic bivalve genera have been reported so far for the Early Jurassic.

Interestingly enough, the quantitative analysis suggests that by the Pliensbachian-Toarcian interval there was no clear differentiation of biochoremas yet. Cluster analysis shows one major group with similarity values higher than 0.50, encompassing South America, the Himalayas, Madagascar, and Iran, with South American localities more related to Madagascar.

Furthermore, BSN shows no clear pattern either, with all regions presenting high similarities with Central Argentina and Chile (Fig. 5.7a). This phenomenon has two possible explanations: on one hand, a sampling bias cannot be disregarded: in a better

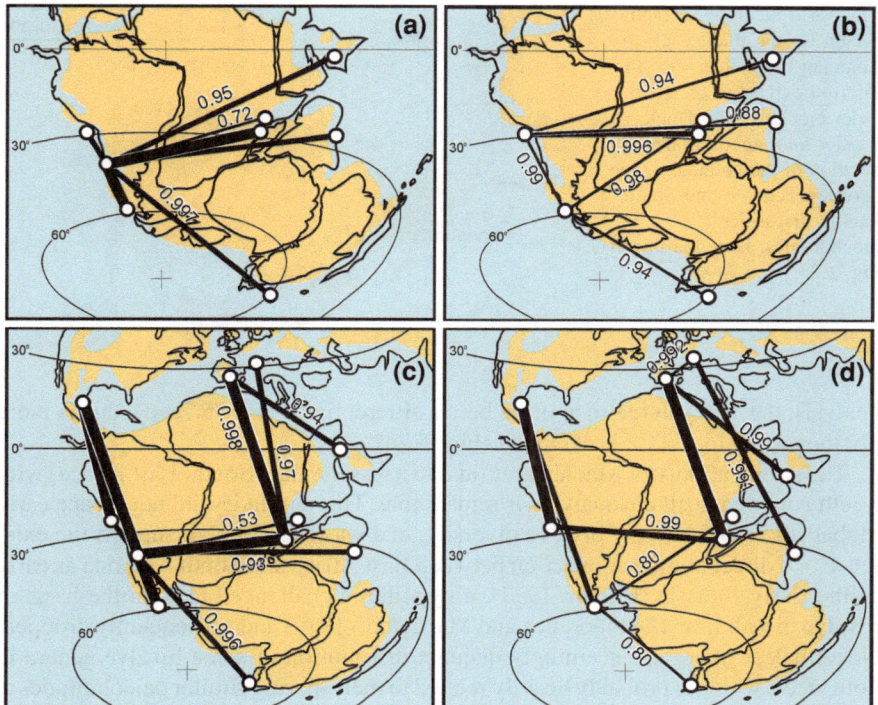

Fig. 5.7 Four BSN analyses of South Hemisphere regions for the Pliensbachian-Toarcian interval. **a** Including all Southern Hemisphere regions with enough data, **b** Excluding region 3 (central Argentina and Chile) from the analysis, **c** Including all Southern Hemisphere regions plus three regions situated in the Paleo-Northern Hemisphere: Mexico, Iberian Peninsula, and southern France, **d** Same as c excluding region 3. Continuous thick lines: distance value <0.10 (more than 90 % of genera shared); continuous lines: distance value ≥0.10 and <0.25 (90 % to more than 75 % of genera shared); continuous thin lines ≥0.25 and <0.50 (75 % to more than 50 % of genera shared); support values of 1 unless otherwise indicated on each line

known region there may be higher chances of finding genera shared with other regions. Table 5.2 shows that South America (especially region 3) is by far the area with more known genera. On the other hand, the results of the BSN may be showing an actual pattern, consequence of a major biogeographic event: the aperture of the Hispanic Corridor. The development of this migration route may have allowed for faunal mixing among some regions; if this were the case, then those regions should be highly similar to all other regions since it would be lumping genera common to both main realms. Both of these explanations may be responsible for the pattern observed here; thus, another BSN was built excluding central Argentina and Chile, resulting in the scheme of Fig. 5.7b. As can be seen dissimilarity is still low (although higher than in Fig. 5.7a), but an incipient differentiation can be seen between northern and southern regions. When some Northern Hemisphere localities are added to the

Fig. 5.8 Paleobiogeographic sketch for the Pliensbachian-Toarcian time slice, showing inferred extension of second-order biochoremas. Thick *broken line*: approximate position of the boundary zone between Tethyan and South Pacific first-order biochoremas. Base map as in Fig. 1.2

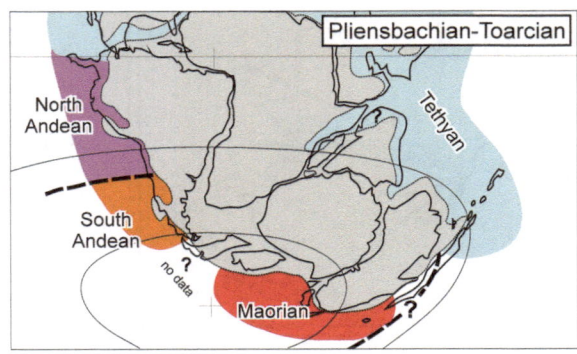

analysis, the Tethyan realm appears better differentiated (Fig. 5.7c–d); this is more obvious when locality 3 is excluded (Fig. 5.7d).

The strong affinity of Madagascar (and to a lesser extent South–East Africa) with South America in all the analyses is remarkable. This similarity did not escape early global studies of Jurassic bivalve diversity, and was even used to support the existence of a direct marine connection between East Africa and South America as early as the Early Jurassic (Hallam 1977), a hypothesis challenged also on the basis of bivalve distribution (Damborenea and Manceñido 1979) and subsequently dropped. Nevertheless, there is a seemingly disjunct distribution of some bivalve genera in both areas, which is probably heavily related to their sharing similar paleolatitudes at the time. Although initial rifting between eastern and western Gondwana probably began late in the Early Jurassic, actual sea-floors spreading apparently started only in the Late Jurassic (Crame 1999), which is in agreement with our results (see discussion below under "Early Cretaceous"). Figure 5.8 shows a graphic summary of the Pliensbachian-Toarcian paleobiogeographic units in the Southern Hemisphere, with the hypothetical paleoposition of the boundary zones.

5.3.3 Middle Jurassic

During Middle Jurassic (Fig. 5.9) and early Late Jurassic times the Southern Hemisphere bivalve benthonic faunas maintained their high proportion of cosmopolitan genera, always above 40 % of all taxa, but still some endemic high-latitude genera were present.

The two South Pacific subunits maintain their level of endemism between 16 and 29 % for the whole Middle Jurassic (see Table 5.3), although at the same time they experienced the proportional decrease of high-latitude taxa. The influx of high-latitude genera reached 24 % for the Maorian and 33 % for the South Andean units, and in the Maorian biochorema it was exceeded during the Late Jurassic. On this evidence, many authors (see Grant-Mackie et al. 2000) concluded that the Maorian Province ceased to exist in the early Middle Jurassic. Nevertheless, a

Fig. 5.9 Paleobiogeographic
sketch for the Aalenian-
Bajocian time slice, showing
inferred extension of second-
order biochoremas. Thick
broken line: approximate
position of the boundary zone
between Tethyan and South
Pacific first-order
biochoremas. Base map as in
Fig. 1.2

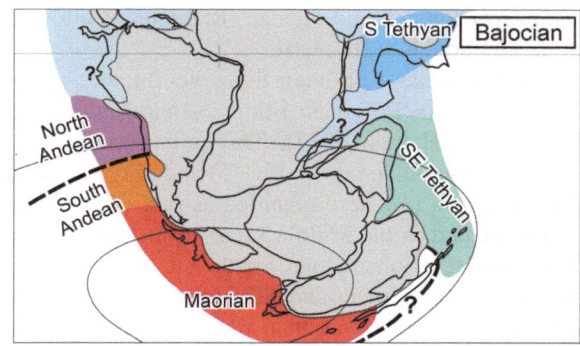

paleogeographic unit is still recognizable (though probably with lower rank)
throughout the Middle Jurassic (Damborenea 2002b). This unit can either be
regarded as an extension of the Maorian Province, or as the origin of the Austral
Province recognized by Kauffman (1973) for Early Cretaceous times. To name it for
the Middle Jurassic the first alternative was followed by Damborenea (2002b), but
regardless of how we chose to call it, the continuity of an austral biogeographic unit
from Triassic (Maorian Province) to Cretaceous (Austral Province) is evident from
our results. This unit is recognized on the basis of the following endemic genera:
Haastina Marwick (Bajocian), *Malagasitrigonia* Cox (Aalenian-Bajocian), *Kanakimya* Campbell and Grant-Mackie (Aalenian-Bajocian?), and *Moewakamya*
Campbell and Grant-Mackie (Bathonian-Oxfordian). For the South Andean unit
endemic genera include *Anditrigonia* Levy (Bajocian-Tithonian), *Andivaugonia*
Leanza (Bajocian-Callovian), *Eoanditrigonia* Leanza (Bajocian-Callovian), *Neuquenitrigonia* Leanza and Garate (Aalenian-Bajocian), *Lambertrigonia* Leanza
(Callovian). High-latitude taxa in these two units are proportionately fewer; they
include *Retroceramus* Koshelkina (Bajocian-Oxfordian) and *Scaphogonia* Crick-
may (Bajocian-Tithonian) for both units, *Praebuchia*? (Aalenian) for the South
Andean, and *Fractoceramus* Koshelkina (Bajocian) and *Hijitrigonia* Kobayashi
(Bajocian) for the Maorian unit. Despite the early Toarcian extinction event,
endemism in the Maorian unit climbed from 22 % in the late Early Jurassic to 28 %
in the early Middle Jurassic. In this connection, in the Andean region endemic
bivalve species were more affected than others by the "Pliensbachian-Toarcian"
extinction event (Aberhan and Fürsich 2000).

The North Andean endemic center is still faintly distinct during the Aalenian,
with the presence of the endemic *Gervilleiognoma* Aberhan and Hillebrandt.
Published data from younger times in this area appear very scarce and incomplete,
thus precluding confident use in this context.

A new SE Tethyan unit (Fig. 5.9) appears to be recognizable from Bajocian
times onwards on the basis of its edemic genera *Indolucina* Fürsich et al.
(Bajocian-Oxfordian), *Agrawalimya* Singh et al. (Bathonian-Callovian), *Indocor-
bula* Fürsich et al. (Bathonian-Callovian), *Indomya* Jaitly (Bathonian-Callovian),

Indoweyla Fürsich and Heinze (Bathonian-Callovian), and *Venilicyprina* Fürsich et al. (Bathonian-Callovian).

For Bathonian-Callovian times yet another new unit began to be discernible: East African (= Provincia Etiopico-Indo-Malgascia in Ficcarelli 1968; Ethiopian Province in Hallam 1977). This unit contains, however, a high percentage of strictly low-latitude genera (82 %) and only one probably high-latitude genus. For these reasons, it is here regarded as part of the Tethyan Realm at this time.

By the end of the Middle Jurassic, just four biochoremas are identifiable in the Southern Hemisphere, two South Pacific (Austral and South Andean), and two Tethyan (East African and SE Tethyan) units. The South Tethyan unit was still recognizable but it was then placed in the Northern Hemisphere as a result of tectonic plate migration to the north.

5.3.4 Late Jurassic

The broad situation observed for Bathonian-Callovian times is generally maintained for the Late Jurassic (Fig. 5.10). The Maorian unit is extended geographically to include the Antarctic and South Atlantic Plateau regions, and even some western Pacific localities (Timor, Sula, Buru, Ceram, see Hayami 1984), but is somewhat less strongly recognizable, and its rank should be lowered, since its overall endemism at the generic level diminishes to 18 % during Oxfordian-Kimmeridgian and to only 14 % during Tithonian-Berriasian times. Endemic genera include *Moewakamya* (Oxfordian), *Jeletzkiella* (Oxfordian-Kimmeridgian), and *Praeaucellina* (Tithonian-Berriasian), while *Malayomaorica* (Oxfordian-Kimmeridgian) has an austral distribution reaching Australia-New Guinea. *Retroceramus*, *Scaphogonia*, *Lyapinella*, and *Anopaea* are high-latitude genera, mostly didemic in distribution.

On the other hand, the South Andean unit maintained its levels of endemism. The anomalous very high percentage of endemism for the Oxfordian-Kimmeridgian is due to the small number of taxa reported so far, and this figure is not statistically significant. Endemic genera include the trigonioideans *Anditrigonia* (Bajocian-Tithonian), *Antutrigonia* (Tithonian-Berriasian), *Notoscabrotrigonia* (Tithonian), and *Splenditrigonia* (Tithonian-Berriasian), and *Retroceramus* and *Scaphogonia* represent the high-latitude taxa.

The East African unit maintained a percentage of endemic taxa between 13 and 21 %, including *Africomiodon* and *Eoseebachia* (Oxfordian-Kimmeridgian), with a very high percentage (just over 70 %) of low-latitude taxa. It is interesting to note here that the influence of high-latitude taxa in this area increased from 2 % in the Bathonian-Callovian to 9 % in the Oxfordian-Kimmeridgian.

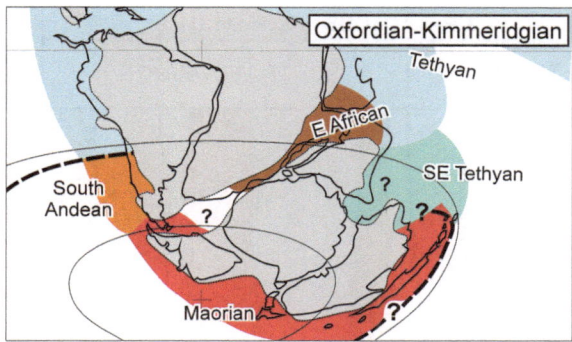

Fig. 5.10 Paleobiogeographic sketch for the Oxfordian-Kimmeridgian time slice showing inferred extension of second-order biochoremas. Thick *broken line*: approximate position of the boundary zone between Tethyan and South Pacific first-order biochoremas. Base map as in Fig. 1.2

5.3.5 *Early Cretaceous*

By the end of the Jurassic and beginning of the Cretaceous the situation changed and antitropical bivalve genera were again well established at least within inoceramoids and monotoids (Crame 1986, 1993; Dhondt 1992). According to Kauffman (1973), recognition of a relatively mature "South Temperate Realm" is clear, with well-defined "Austral" and "East African" provinces as subordinate biogeographic units by the beginning of the Cretaceous. The change of the East African unit from the Tethyan (during most of the Jurassic) to the South Pacific realm is already hinted by the relatively high percentage (9 %) of high-latitude taxa during Oxfordian-Kimmeridgian times, possibly as a direct consequence of the opening of the Mozambique corridor or trans-Gondwana seaway, which communicated this part of Africa with the South Pacific (see discussion in Crame 1999). Kauffman (1973) referred the East African province to the Indo-Pacific Region in the Early Cretaceous, according to its content of southern Pacific bivalve lineages, with a geographic range now restricted to Southern Africa, Madagascar, and Tanzania. To the north, a transition zone developed (India, Arabia, NE Africa) as continuation of the Jurassic SE Tethyan biochorema, with strong Tethyan influence (i.e., the north Indian Ocean subprovince in Kauffman 1973).

During the Early Cretaceous (Berriasian) the South Andean unit contained a few endemic trigoniodeans, such as *Antutrigonia*, *Splenditrigonia*, and *Transitrigonia*. *Anopaea* was a genus with didemic distribution, and during the Berriasian it was present in New Zealand and Antarctica, while the endemic *Praeaucellina* lingered from Tithonian times in the Maorian unit. According to Kauffman (1973), the Austral Province of the Indo-Pacific Region (his South Temperate Realm) was strongly developed at the beginning of the Cretaceous, including Australia, New Zealand and New Guinea.

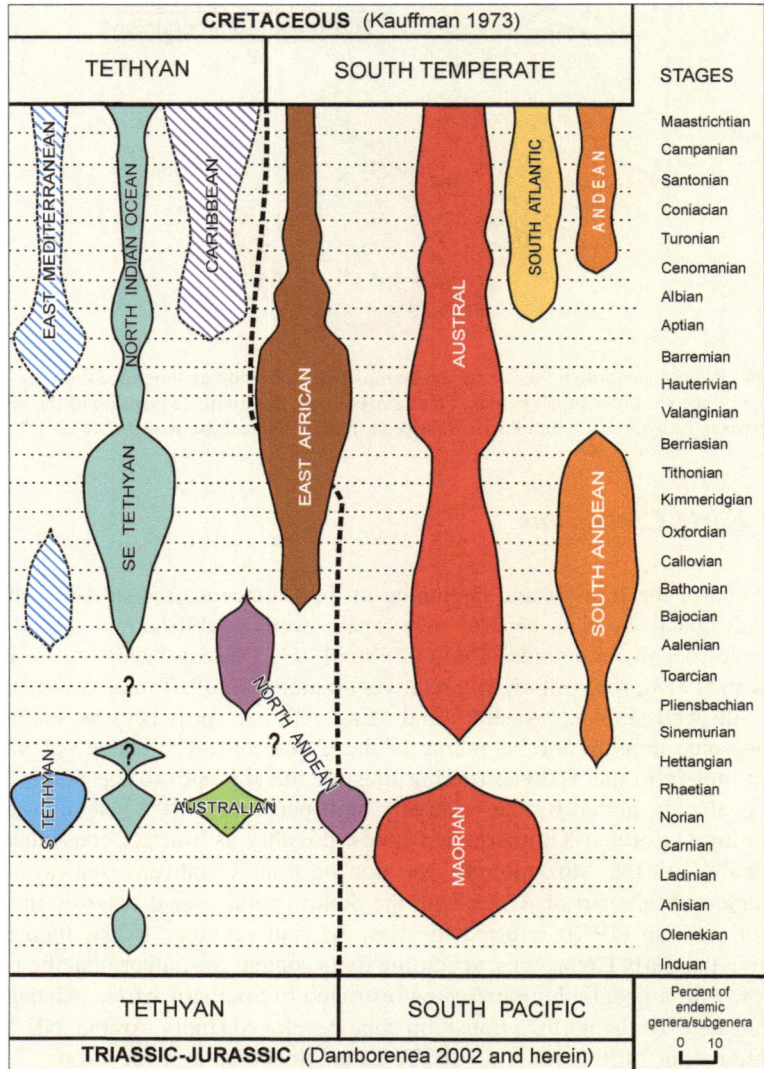

Fig. 5.11 Evolution of the second-order paleobiogeographic units discussed here, recognized for the Southern Hemisphere using marine benthonic bivalve faunas. Triassic and Jurassic data discussed here, Cretaceous data from Kauffman (1973). The width of the bars is proportional to the percentage of endemic genera through time, which is an indication of their change in rank. Solid color indicates development in the Paleo-Southern Hemisphere; diagonal pattern shows units which migrated to the Paleo-Northern Hemisphere (undated from Damborenea 2002c, Fig. 7)

From Middle Cretaceous times onwards the North Andean unit was clearly recognizable as a Caribbean subprovince/province (Kauffman 1973), and an Austral Realm was proposed by Fleming (1963) on the basis of bivalve and gastropod distribution.

Integrating our results (Damborenea 2002b, Fig. 6, updated here) with those outlined by Kauffman (1973), the overall history of the paleobiogeographic units just discussed during Triassic to Cretaceous times shows some interesting features (Fig. 5.11). The disruption of the pre-existing pattern at the Triassic/Jurassic boundary, coinciding with the severe biotic crisis, is evident, as well as the relatively rapid recovery. Apart from this post-extinction period, the Maorian-Austral unit shows a remarkable continuity in time and may be regarded as precursor of the Late Cretaceous-Early Cenozoic Weddellian Province (Crame 1999).

If a ranking related to degree of endemism is used, it is evident that units frequently changed rank (represented by the changing width of their graphic representation on Fig. 5.11). Moreover, it is clear that some units changed their relation to higher order units, too. This happened with the East African unit, which first developed as part of the Tethyan Realm but toward the end of the Jurassic began to receive more influence from southern faunas to eventually be regarded as part of the South Temperate Realm by Kauffman (1973) in the Cretaceous.

5.4 Congruence

Since this analysis was only based on one group of organisms, it is interesting to compare the results with those obtained from other groups. Different groups of organisms sometimes show congruent patterns of biogeographic distribution, and these are extremely important to the recognition of general patterns which could be related to common causes. Nevertheless, most other benthonic macro-invertebrate groups (gastropods, corals and echinoderms) are still insufficiently known to be used in paleobiogeography for the Jurassic at a global scale. The only groups with comparable level of analysis for the time considered here are ammonites and belemnites, but they had mostly nektonic or nekto-benthonic habits and thus their distribution patterns may be partly influenced by somewhat different phenomena.

From the scattered available data it is evident that some groups of restricted tropical affinities, such as reef building corals and certain sponges like *Stylothalamia*, are completely absent from the Southern Hemisphere low-latitude areas belonging to the South Pacific paleobiogeographic unit as understood here (Hillebrandt 1981; Crame 1987; Beauvais 1992).

One of the most interesting questions that need further discussion is the observed differences in the first-order paleobiogeographic units for the Southern Hemisphere between those based on bivalves as described here and those based on ammonites. Differences between distribution of benthonic and pelagic organisms are to be expected (see discussion in Masse 1992). According to the pattern of ammonite distribution most authors consistently recognize only two Realms

(Boreal and Tethyan) for the Jurassic (see Arkell 1956, and more references in Cecca 1999; Westermann 2000b; Grant-Mackie et al. 2000).

Hillebrandt (1981; Hillebrandt et al. 1992) distinguished an Eastern Pacific and an Indo-SW Pacific sub-realm during some time lapses within the Early and Middle Jurassic, but they admit that for the Late Jurassic lack of data severely constrains the discussion of ammonite provincialism in circum-Pacific regions. Dommergues et al. (2001) reviewed the distribution of Early Jurassic ammonoids trying to match different morphologic sets previously recognized with analytic distribution patterns, defined by considering the distribution and abundance of species, but they did not use data from the southern high-paleolatitude regions. In fact, a different picture seems to be emerging as new data become known.

Westermann (1981) recognized an East Pacific ammonite realm (sub-realm in Westermann and Hudson 1991) for late Bajocian to early Callovian times extending from Arctic Canada along the Pacific coasts of America and west Antarctica to New Zealand. Riccardi (1991) found a nearly continuous presence of endemic ammonites from Pliensbachian to Oxfordian in South America, with a sharp increase during Aalenian, and almost no endemics during Kimmeridgian-Berriasian. Westermann (1996) recognized a high proportion of endemic taxa for the Bajocian of New Zealand, with a marked increase in Andean affinities for the Bathonian-Callovian (see Grant-Mackie et al. 2000 for references). According to Enay and Cariou (1997), an austral ammonite fauna of low diversity became progressively better established around east and south Gondwana from Oxfordian times onwards.

An Indo-Pacific paleobiogeographic unit was recognized on the basis of belemnite distribution from Callovian times onwards, initially given realm rank (Stevens 1963), though linked to (or derived from) the Tethys, but only Boreal and Tethyan Realms were later accepted (Stevens 1973). The contrast between the sharp and stable Boreal/Tethyan boundary and a diffuse and unstable Tethyan/Indo-Pacific boundary of belemnite paleobiogeographic units is emphasized by all authors. According to Challinor (1991) and Challinor et al. (1992) a South Pacific Province (referred to the Tethyan Realm) may be distinguished during the Jurassic. This faunal province extended along the coast of Gondwanaland from southern South America to New Zealand and possibly New Caledonia in the Jurassic (Challinor and Hikuroa 2007). Belemnitina (which defines the Boreal Realm) was dominant in Boreal regions throughout the Jurassic and was also known in South America-New Zealand-New Caledonia up to the Middle Jurassic (Challinor et al. 1992; Doyle et al. 1997). During the Middle and Late Jurassic trans-Gondwana migrations introduced Tethyan elements to southern South America and Antarctica (Challinor and Hikuroa 2007). In the Jurassic there was no belemnite species in common between Tethyan and Pacific coasts of Gondwanaland, and two provinces are recognized, respectively; Indo-Tethyan and South Pacific (Challinor in Grant-Mackie et al. 2000).

Gastropods have been used to recognize paleobiogeographic units for the Cretaceous (Sohl 1987), but Triassic and Jurassic gastropods are poorly known at a global scale yet. Within these limitations, Tong and Erwin (2001) noted that the relationships between Triassic Tethyan and American gastropods are weak, even at the generic level.

Triassic and Early Jurassic brachiopods from austral regions do show a considerable degree of endemism which supports the recognition of a Maorian Province (MacFarlan 1992; Manceñido and Dagys 1992; Manceñido 2002 and references therein). During the Late Triassic a strong Maorian unit was well characterized by (among others) the spiriferide *Rastelligera*, the athyride *Clavigera*, and the rhynchonellide *Sakawairhynchia*, the last one widely distributed in East Asia, New Zealand, New Caledonia (MacFarlan in Grant-Mackie et al. 2000), and west central Argentina (Damborenea and Manceñido 2012). Close similarities of some rhynchonellides and spiriferinides from New Zealand and South America persist into the Early Jurassic but austral endemism was lower, though there were probably also some bipolar genera. Maorian brachiopod endemism extended during Middle and Late Jurassic in New Zealand-New Caledonia (MacFarlan in Grant-Mackie et al. 2000). Brachiopod diversity diminished considerably toward the Late Jurassic in the whole austral regions.

Knowledge of Jurassic radiolarians from the Southern Hemisphere is just emerging (see Kiessling and Scasso 1996; Aita et al. in Grant-Mackie et al. 2000), but clearly points to non-Tethyan affinities for the Late Jurassic faunas from Antarctica and the Waipapa Terrane (New Zealand). These data support a paleolatitudinal zonation for Late Jurassic radiolarian faunas in the Southern Hemisphere, with a distinct austral paleobiogeographic unit (which is not symmetric with the boreal one), characterized by low endemism and dominated by *Parvicingula-Praeparvicungula* (Kiessling and Scasso 1996).

Other microfossils which have been discussed in this context are ostracods (Boomer and Ballent 1996; Mette 2004; Arias 2006) and interesting results are emerging, though information is still too fragmentary to show comparable detail with extensive data from bivalves.

Brief references to paleoclimate and paleogeography can be made to help explain these differences. During the Jurassic, climates are said to have been milder, with temperatures more equable across latitude, but with some seasonality due to the different day lengths through the year (see Sect. 1.5). This may have been a particularly important factor affecting marine benthonic biota at high latitudes. Pelagic taxa are probably less affected by seasonality than benthonic inhabitants of littoral environments, such as bivalves. But apart from climate, purely geographic factors, affecting the ocean circulation patterns, may have had considerable influence, too. For instance, the distribution of land and sea during the Jurassic in high latitude regions: while in boreal regions there was a polar ocean, nearly completely surrounded by land or shallow seas, the austral regions were part of the proto-Pacific ocean, which was continuous with the eastern Tethys. Enay and Cariou (1997) emphasized that the lack of such geographic trap in austral regions would explain why the austral ammonite faunas were never as distinct from the Tethyan as the Boreal ones. In addition, an important migration route for benthonic organisms since the Early Jurassic, the Hispanic Corridor (Damborenea and Manceñido 1979; Smith 1989; Boomer and Ballent 1996; etc.), remained as a barrier for most pelagic animals until late Middle or even Late Jurassic times (Elmi 1993; Damborenea 2000; Aberhan 2001).

Regarding the evidence from terrestrial organisms, Balme (in Grant-Mackie et al. 2000) indicated that there is general agreement that latitudinal zonations based on megafossil plant distribution existed in the Jurassic, but these do not support the existence of any substantial climatic barriers.

5.5 Paleobiogeographic Units and Mass Extinctions

The largest biotic crises did not only affect faunal taxonomic diversity, but also had serious impacts on other aspects as well, the ecologic one being the most well known. Several authors had also emphasized that one of the most direct results of such severe taxonomic and ecologic disruption is the loss of previous paleobiogeographic structures (see discussion and references in Ros and Echevarría 2011). It is well-known that endemic taxa seem to be positively selected by extinction and suffer more than widely distributed taxa (Hallam and Miller 1988). As the recognition of paleobiogeographic structures is largely based on the presence of endemic taxa, the disruption of the previous pattern is the most likely result.

As expected, the then existing late Paleozoic paleobiogeographic patterns were seriously disrupted by the Permo-Triassic extinction event. This has been already noticed by previous authors, and it is clearly seen in Shi and Grunt 2000 (Fig. 8), where the Permian paleobiogeographic regions recognized for the Anti-Boreal (or Gondwana) areas disappeared altogether at the beginning of the Triassic. Our analysis shows that there was no recognizable paleobiogeographic pattern for all Early and part of Middle Triassic times, and endemism in the Southern Hemisphere re-appeared only by Etalian times (i.e., late Anisian) in New Zealand. It became stronger by the Ladinian, when already a Maorian unit was well characterized, and was definitively settled during the late Triassic, when other endemic centers also appeared (see Fig. 5.11).

This apparently well-established late Triassic paleobiogeographic pattern, with at least five biochoremas of various ranks, was again almost totally wiped out by the next biotic crisis, the Triassic-Jurassic extinction event. The earliest Jurassic low endemism level was not restricted to the southern hemisphere, but is also apparent at a global scale: Liu (1995) concluded that provinciality in the northern hemisphere during Hettangian, Sinemurian and even Pliensbachian times was unrecognizable. Our data clearly show this pattern for the end Triassic extinction, and it is also evident that the biota and endemic centers took some time to recover (see discussion in Hallam 1996), but recovery appears to have been somewhat faster than after the Permo-Triassic event. This is particularly clear in the previously well-defined Maorian biochorema, as the lack of endemic genera makes this unit impossible to recognize during Hettangian and Sinemurian times, though it recovered by Pliensbachian times.

Macchioni and Cecca (2002) register two drops in ammonite diversity in the Early Toarcian of Europe, the first one producing the disruption of Tethyan-Boreal provinciality, the second one coinciding with the onset of OAE. They further proposed that trends in endemism may be reversed during transgressions or regressions. Our data are not refined enough to pick any consequence of the Early Toarcian biotic crises which is known to have affected bivalve biodiversity worldwide.

Another quite different but interesting relationship between biogeography and extinction events is just now emerging. The new developments in paleobiogeographic analysis suggest that extinction may negatively influence the retrieval of biogeographic patterns from living organisms (Lieberman 2003). For this reason, the careful analysis of distribution of extinct taxa acquires an extra interest.

References

Abbate E, Ficcarelli G, Pirini Radrizzani C, Salvietti A, Torre D, Turi A (1974) Jurassic sequences from the Somali coast of the Gulf of Aden. Riv Ital Paleontol Stratigr 80:409–478

Aberhan M (1994) Early Jurassic Bivalvia of northern Chile. Part I. Subclasses Palaeotaxodonta, Pteriomorphia, and Isofilibranchia. Beringeria 13:1–115

Aberhan M (2001) Bivalve palaeobiogeography and the Hispanic Corridor: time of opening and effectiveness of a proto-Atlantic seaway. Palaeogeogr Palaeoclimatol Palaeoecol 165:375–394

Aberhan M (2004) Early Jurassic Bivalvia of northern Chile. Part II. Subclass Anomalodesmata. Beringeria 34:117–154

Aberhan M (2007) A new species of *Coelastarte* (Astartidae: Bivalvia) from the Early Jurassic of Chile and unusual ornamentation in eastern Pacific Jurassic bivalves. Beringeria 37:3–9

Aberhan M, Fürsich FT (2000) Mass origination versus mass extinction: the biological contribution to the Pliensbachian-Toarcian extinction event. J Geol Soc Lond 157:55–60

Aberhan M, Hillebrandt A (1996) Taxonomy, ecology, and palaeobiogeography of *Gervilleioperna* (*Gervilleiognoma*) *aurita* n. subgen. n. sp. (Bivalvia) from the Middle Jurassic of northern Chile. Paläontol Z 70:79–96

Aberhan M, Hillebrandt A (1999) The bivalve *Opisoma* in the Lower Jurassic of northern Chile. Profil 16:149–164

Aberhan M, Bussert R, Heinrich WD, Schrank E, Schultka S, Sames B, Kriwet J, Kapilima S (2002) Palaeoecology and depositional environments of the Tendaguru Beds (Late Jurassic to Early Cretaceous, Tanzania). Mitt Mus Natkd BerlIN, Geowiss Reihe 5:19–44

Ager DV (1975) The Jurassic World Ocean (with special reference to the North Atlantic). In: Finstad KG, Selley RC (eds) Jurassic Northern North Sea Symposium. North Sea Petrol Soc, Oslo:1–43

Agrawal SK (1956a) Deux lamellibranches intéressants de la Série de Katrol dans le Kutch (Inde). Bull Soc Géol Fr Ser 6:13–19

Agrawal SK (1956b) Contribution à l'étude stratigraphique et paléontologique du Jurassique du Kutch (Inde). Ann Centre Étud Document Paléontol 19:1–188

Agrawal SK, Rai JN (1978) On a new species of *Prorokia* Böhm from the Jurassic of Kutch (Gujarat, India). J Geol Soc India 19:373–375

Arias C (2006) Northern and Southern Hemispheres ostracod palaeobiogeography during the Early Jurassic: Possible migration routes. Palaeogeogr Palaeoclimatol Palaeoecol 233:63–95

Arkell WJ (1956) Jurassic geology of the world. Oliver and Boyd Ltd, Edinburgh–London

Avias J (1953) Contribution a l'étude stratigraphique et paléontologique des formations antécrétacées de la Nouvelle Calédonie centrale. Sci Terre 1:1–276

Barrabé L (1929) Contribution a l'étude stratigraphique et pétrographique de la partie médiane du pays Sakalave (Madagascar). Mém Soc Géol France NS 5:1–270

Basse E (1930) Contribution a l'etude du Jurassique superieur (facies Corallien) en Ethiopie et en Arabie meridionale. Mem Soc Geol France NS 14:105–155

Beauvais L (1992) Corals of the circum-Pacific region. In: Westermann GEG (ed) The Jurassic of the circum-Pacific. Cambridge University Press, New York

Begg JG, Campbell HJ (1985) *Etalia*, a new Middle Triassic (Anisian) bivalve from New Zealand, and its relationship with other pteriomorphs. NZ J Geol Geophys 28:725–741

Besairie H (1930) Recherches géologiques à Madagascar. Contribution à l'étude des ressources minérales. Impr. H. Basuyau & Cie, Toulouse

Besairie H, Collignon H (1972) Géologie de Madagascar. I, Les terrains sédimentaires. Ann Géol Madagascar 35:1–463

Boomer I, Ballent S (1996) Early-Middle Jurassic ostracod migration between the northern and southern hemispheres: further evidence for a proto Atlantic-Central America connection. Palaeogeogr Palaeoclimatol Palaeoecol 121:53–64

Burckhardt C (1900a) Profils géologiques transversaux de la Cordillère Argentino-Chilienne. Stratigraphie et tectonique. An Mus La Plata, Sec Geol Mineral 2:1–136

Burckhardt C (1900b) Coupe géologique de la Cordillère entre Las Lajas et Curacautin. An Mus La Plata, Sec Geol Mineral 3:1–100

Burckhardt C (1903) Beiträge zur Kenntnis der Jura- und Kreideformation der Cordillere. Palaeontographica 50:1–144

Bürgl H (1961) El Jurásico e Infracretáceo del río Batá, Boyacá. Inst Geol Nac Bogotá Bol Geol 6:169–211

Bussert R, Heinrich WD, Aberhan M (2009) The Tendaguru Formation (Late Jurassic to Early Cretaceous, southern Tanzania): definition, palaeoenvironments, and sequence stratigraphy. Fossil Record 12:141–174

Cecca F (1999) Palaeobiogeography of Tethyan ammonites during the Tithonian (latest Jurassic). Palaeogeogr Palaeoclimatol Palaeoecol 147:1–37

Cecca F, Westermann GEG (2003) Towards a guide to palaeobiogeographic classification. Palaeogeogr Palaeoclimatol Palaeoecol 201:179–181

Cecioni G, Westermann GEG (1968) The Triassic/Jurassic marine transition of coastal Central Chile. Pacific Geol 1:41–75

Challinor AB (1991) Belemnite succession and faunal provinces in the Southwest Pacific, and the belemnites of Gondwana. J Austral Geol Geophys 12:301–325

Challinor AB, Hikuroa DCH (2007) New middle and upper Jurassic belemnite assemblages from West Antarctica (Latady Group, Ellsworth Land): taxonomy and paleobiogeography. Palaeontol Electron 10 2.8 MB; http://palaeo-electronica.org/paleo/2007_1/assemblage/index.html

Challinor AB, Doyle P, Howlett PJ, Nal'nyaeva TI (1992) Belemnites of the circum-Pacific region. In: Westermann GEG (ed) The Jurassic of the circum-Pacific. Cambridge University Press, New York

Chong DG, Hillebrandt A (1985) El Triásico preandino de Chile entre los 23° 30' y 26° 00' de lat. Sur. Actas 4° Congr Geol Chileno (Antofagasta) 1:162–210

Coleman PJ, Skwarko SK (1967) Lower Triassic and Middle Jurasic fossils at Enanty Hill, Mingenew, Perth Basin, West Australia. Bull Bur Min Res. Geol Geophys (Australia) 92:197–214

Cox LR (1935a) Jurassic Gastropoda and Lamellibranchia. In: MacFadyen W et al (eds) The Mesozoic palaeontology of British Somaliland (Part II of the Geology and Palaeontology of British Somaliland) 8:148–197

Cox LR (1935b) The Triassic, Jurassic and Cretaceous Gastropoda and Lamellibranchia of the Attock District. Mem Geol Surv India Paleontologia Indica NS 20:1–27

Cox LR (1936) Fossil Mollusca from Southern Persia (Iran) and Bahrein Island. Mem Geol Surv India Paleontologia Indica NS 22:1–69

Cox LR (1940) The Jurassic lamellibranch fauna of Kuchh (Cutch). Mem Geol Surv India Palaeontologia Indica, Ser 9, 3, 3:1–157

Cox LR (1949) Moluscos del Triásico superior del Perú (Upper Triassic Mollusca from Perú). Bol Inst Geol Perú 12:1–50

Cox LR (1952) The Jurassic lamellibranch fauna of Cutch (Kachh). No. 3, families Pectinidae, Amusiidae, Plicatulidae, Limidae, Ostreidae and Trigoniidae (Supplement). Mem Geol Surv India. Palaeontologia Indica, Ser 9, 3, 4:1–128

Cox LR (1956) Jurassic Mollusca from Peru. J Paleontol 30:1179–1186

Cox LR (1965) Jurassic Bivalvia and Gastropoda from Tanganyika and Kenya. Bull Brit Mus (Nat Hist). Geol Suppl 1:1–213

Crame JA (1981) The occurrence of Anopaea (Bivalvia: Inoceramidae) in the Antarctic Peninsula. J Mollusc Stud 47:206–219

Crame JA (1982a) Late Mesozoic bivalve biostratigraphy of the Antarctic Peninsula region. J Geol Soc Lond 139:771–778

Crame JA (1982b) Late Jurassic inoceramid bivalves from the Antarctic Peninsula and their stratigraphic use. Palaeontology 25:555–603

Crame JA (1983) The occurrence of the upper Jurassic bivalve Malayomaorica malayomaorica (Krumbeck) on the Orville Coast, Antarctica. J Mollusc Stud 49:61–76

Crame JA (1984) Preliminary bivalve zonation of the Jurassic-Cretaceous boundary in Antarctica. Mem 3° Congr Latinoam Paleontol (Mexico), UNAM: 242–254

Crame JA (1985) Lower Cretaceous inoceramid bivalves from the Antarctic Peninsula region. Palaeontology 28:475–525

Crame JA (1986) Late Mesozoic bipolar bivalve faunas. Geol Mag 123:611–618

Crame JA (1987) Late Mesozoic bivalve biogeography of Antarctica. Proc Sixth Gondwana Symp (Columbus, Ohio): 93–102

Crame JA (1992) Evolutionary history of the Polar Regions. Hist Biol 6:37–60

Crame JA (1993) Bipolar molluscs and their evolutionary implications. J Biogeogr 20:145–161

Crame JA (1996a) Evolution of high-latitude molluscan faunas. In: Taylor JD (ed) Origin and evolutionary radiation of the Mollusca. Oxford University Press, Oxford

Crame JA (1996b) Antarctica and the evolution of taxonomic diversity gradients in the marine realm. Terra Antarct 3:121–134

Crame JA (1996c) A new oxytomid bivalve from the Upper Jurassic-Lower Cretaceous of Antarctica. Palaeontology 39:615–628

Crame JA (1997) An evolutionary framework for the Polar Regions. J Biogeogr 24:1–9

Crame JA (1999) An evolutionary perspective on marine faunal connections between southernmost South America and Antarctica. Scientia Marina 53(Supp 1):1–14

Crame JA (2000a) Evolution of taxonomic diversity gradients in the marine realm: evidence from the composition of recent bivalve faunas. Paleobiology 26:188–214

Crame JA (2000b) The nature and origin of taxonomic diversity gradients in marine bivalves. In: Harper EM, Taylor JD, Crame JA (eds) The evolutionary biology of the Bivalvia. Geol Soc Spec Publ 177:347–360

Crame JA, Kelly AR (1995) Composition and distribution of the inoceramid bivalve genus Anopaea. Palaeontology 38:87–103

Crame JA, Pirrie D, Crampton JS, Duane AM (1993) Stratigraphy and regional significance of the Upper Jurassic-Lower Cretaceous Byers Group, Livingston Island, Antarctica. J Geol Soc Lond 15:1075–1087

Crampton JS (1988) Comparative anatomy of the bivalve families Isognomonidae, Inoceramidae, and Retroceramidae. Paleontology 31:965–996

Dacque E (1905) Beiträge zur Geologie des Somalilandes, 2. Teil. Oberer Jura. Beitr Paläontol Geol Öster-Ungarns und des Orients 17:119–160

Damborenea SE (1990) Middle Jurassic inoceramids from Argentina. J Paleontol 64:736–759

Damborenea SE (1993) Early Jurassic South American pectinaceans and circum-Pacific palaeobiogeography. Palaeogeogr Palaeoclimatol Palaeoecol 100:109–123

Damborenea SE (1996) Palaeobiogeography of Early Jurassic bivalves along the southeastern Pacific margin. 13° Congr Geol Argent y 3° Congr Explor Hidrocarb (Buenos Aires) Actas 5:151–167

Damborenea SE (1998) The bipolar bivalve *Kolymonectes* in South America and the diversity of Propeamussiidae in Mesozoic Times. In: Johnston PA, Haggart JW (eds) Bivalves: an eon of evolution—Paleobiological studies Honoring Norman D. Newell University of Calgary Press, Calgary

Damborenea SE (2000) Hispanic Corridor: its evolution and the biogeography of bivalve molluscs. In: Hall RL, Smith PL (eds) Advances in Jurassic research 2000. GeoResearch Forum 6:369–380

Damborenea SE (2002a) Early Jurassic bivalves from Argentina. Part 3: Superfamilies Monotoidea, Pectinoidea. Plicatuloidea and Dimyoidea. Palaeontographica A 265:1–119

Damborenea SE (2002b) Jurassic evolution of Southern Hemisphere marine palaeobiogeographic units based on benthonic bivalves. Geobios 35 MS 24:51–71

Damborenea SE (2002c) Unidades paleobiogeográficas marinas jurásicas basadas sobre moluscos bivalvos: una visión desde el Hemisferio Sur. An Acad Nac Cienc Exact Fís Nat 53:141–160

Damborenea SE (2004) Early Jurassic *Kalentera* (Bivalvia) from Argentina and its palaeobiogeograhical significance. Ameghiniana 41:185–198

Damborenea SE, Lanés S (2007) Early Jurassic shell beds from marginal marine environments in southern Mendoza, Argentina. Palaeogeogr Palaeoclimatol Palaeoecol 250:68–88

Damborenea SE, Manceñido MO (1979) On the palaeogeographical distribution of the pectinid genus *Weyla* (Bivalvia, Lower Jurassic). Palaeogeogr Palaeoclimatol Palaeoecol 27:85–102

Damborenea SE, Manceñido MO (1988) *Weyla*: semblanza de un bivalvo Jurásico andino. Actas 5° Congr Geol Chileno 2:C13–C25

Damborenea SE, Manceñido MO (1992) A comparison of Jurassic marine benthonic faunas from South America and New Zealand. J Roy Soc NZ 22:131–152

Damborenea SE, Manceñido MO (2005) Biofacies analysis of Hettangian-Sinemurian bivalve/brachiopod associations from the Neuquén Basin (Argentina). Geol Acta 3:163–178

Damborenea SE, Manceñido MO (2012) Late Triassic bivalves and brachiopods from southern Mendoza. Rev Paléobiol, VS, Argentina 11

Damborenea SE, Polubotko IV, Sey II, Paraketsov KV (1992) Bivalve zones and assemblages of the circum-Pacific region. In: Westermann GEG (ed) The Jurassic of the circum-Pacific. Cambridge University Press, Cambridge

Dhondt AV (1992) Cretaceous inoceramid biogeography: a review. Palaeogeogr Palaeoclimatol Palaeoecol 92:217–232

Diaz-Romero V (1931) Contributo allo studio della fauna Giurese della Dancalia centrale. Palaeontogr Ital 31:1–61

Diener C (1916) Die marinen Reiche der Trias-Periode. Denks Akad Wissen Wien 92:405–549

Dietrich WO (1933) Zur Stratigraphie und Palaeontologie der Tendaguruschichten. Palaeontographica Suppl 7(2):1–86

Dommergues JL, Laurin B, Meister C (2001) The recovery and radiation of Early Jurassic ammonoids: morphologic versus paleobiogeographical patterns. Palaeogeogr Palaeoclimatol Palaeoecol 165:195–213

Douvillé H (1904) Sur quelques fossiles de Madagascar. Bull Soc Géol France sér 4:207–217

Doyle P (1987) Lower Jurassic—Lower Cretaceous Belemnite biogeography. Palaeogeogr Palaeoclimatol Palaeoecol 61:237–245

Doyle P, Crame JA, Thomson MRA (1990) Late Jurassic-Early Cretaceous macrofossils from Leg 113, Hole 692B, eastern Weddell Sea. Proc Ocean Drill Progr Sci Res 113:443–447

Doyle P, Kelly SRA, Pirrie D, Riccardi AC (1997) Jurassic belemnite distribution patterns: implications of new data from Antarctica and Argentina. Alcheringa 21:219–228

Edwards CW (1980) Early Mesozoic marine fossils from central Alexander Island. Bull Brit Antarct Surv 49:33–58

El-Asa'ad GMA (1989) Callovian colonial corals from the Tuwaiq Mountain Limestone of Saudi Arabia. Palaeontology 32:675–684

Elmi S (1993) Les voies d'échange faunique entre l'Amérique du Sud et la Téthys alpine pendant le Jurassique inférieur et moyen. Doc Lab Géol Lyon 125:139–149

Enay R, Cariou E (1997) Ammonite faunas and palaeobiogeography of the Himalayan belt during the Jurassic: initiation of a late Jurassic austral ammonite fauna. Palaeogeogr Palaeoclimatol Palaeoecol 134:1–38

Etayo-Serna F, Solé de Porta N, De Porta J, Gaona T (2003) The Batá Formation in Colombia is truly Cretaceous, not Jurassic. J S Am Earth Sc 16:113–117

Etheridge J Jr (1910) Oolitic fossils of the Greenough River district, Western Australia. Bull Geol Surv W Australia 36:29–50

Fantini-Sestini N (1966) The geology of the Upper Djadjerud and Lar Valleys (North Iran). II. Palaeontology. Upper Liassic Molluscs from Shemdhak Formation. Riv Ital Paleontol 72:795–852

Feruglio E (1936) Palaeontographia Patagonica. Mem Inst Gel Univ Padova 11:1–384

Ficcarelli G (1968) Fossili giuresi della serie sedimentaria del Nilo Azurro meridionale. Riv Ital Paleontol 74:23–50

Fischer E (1915) Jura und Kreideversteinerungen aus Persien. Beitr Paläontol Geol Österreichs-Ungarns und des Orients 27:207–273

Fleming CA (1959) *Buchia plicata* (Zittel) and its allies, with a description of a new species, *Buchia hochstetteri*. NZ J Geol Geophys 2:889–904

Fleming CA (1962) New Zealand biogeography: a palaeontologist's approach. Tuatara 10:53–108

Fleming CA (1963) The nomenclature of biogeographic elements in the New Zealand Biota. Trans Roy Soc NZ Gen 1:13–22

Fleming CA (1964) History of the bivalve family Trigoniidae in the Southwest Pacific. The geological background to an Australian "living fossil". Austral J Sci 26:196–204

Fleming CA (1987) New Zealand Mesozoic Bivalves of the superfamily Trigoniacea. NZ Geol Surv Palaeontol Bull 53:1–104

Freneix S, Grant-Mackie JA, Lozes J (1974) Présence de *Malayomaorica* (Bivalvia) dans le Jurassique supérieur de la Nouvelle Calédonie. Bull Soc Géol France 16:457–464

Fuenzalida Villegas H (1937) Las capas de Los Molles. Bol Mus Nac Hist Nat (Chile) 16:66–98

Fürsich FT, Heinze M (1998) Contributions to the Jurassic of Kachchh, Western India. VI. The bivalve fauna. Part III. Subclass Palaeoheterodonta (Order Trigonioida). Beringeria 21:151–168

Fürsich FT, Heinze M, Jaitly AK (2000) Contributions to the Jurassic of Kachchh, Western India. VIII. The bivalve fauna. Part IV. Subclass Heterodonta. Beringeria 27:63–146

Fürsich FT, Hautmann M, Senowbari-Daryan B, Seyed-Emami K (2005) The Upper Triassic Nayband and Darkuh formations of east-central Iran: stratigraphy, facies patterns and biota of extensional basins on an accreted terrane. Beringeria 35:53–133

Fütterer K (1897) Beiträge zur Kenntniss des Jura in Ost-Afrika, IV. Der Jura von Schoa (Süd-Abessinien). Z Deut Geol Gesells 49:568–627

Gardner RN (2005) Middle-Late Jurassic bivalves of the superfamily Veneroidea from New Zealand and New Caledonia. NZ J Geol Geophys 48:325–376

Gardner RN (2009) The bivalve genus *Lopatinia* (Family Cucullaeidae) from the Late Jurassic of New Zealand. NZ J Geol Geophys 52:95–99

Gardner RN, Campbell HJ (1997) The bivalve genus *Grammatodon* from the Middle Jurassic of the Catlins District, South Otago, New Zealand. NZ J Geol Geophys 40:487–498

Gardner RN, Campbell HJ (2002) Middle to Late Jurassic bivalves of the genera *Neocrassina* and *Trigonopis* from New Zealand. NZ J Geol Geophys 45:323–347

Gardner RN, Campbell HJ (2007) *Oxyeurax* and *Hemipelex*, new names for *Oxyloma* and *Hemimenia* Gardner & Campbell, 2002, pre-occupied (Mollusca: Bivalvia: Astartidae). NZ J Geol Geophys 50:365

Geiger M, Schweigert G (2006) Toarcian-Kimmeridgian depositional cycles of the south-western Morondova Basin along the rifted continental margin of Madagascar. Facies 52:85–112

Geyer OF (1973) Das präkretazische Mesozoikum von Kolumbien. Geol Jb B 5:1–155

Geyer OF (1977) Die "*Lithiotis*-Kalke" im Bereich der unterjurassischen Tethys. N Jb Geol Paläontol, Abh 153:304–340

Geyer OF (1979) Zur Paläogeographie mesozoischer Ingressionen und Transgressionen in Kolumbien. N Jb Geol Paläontol, Monats 1976(6):349–368

Grant-Mackie JA (1960) On a new *Kalentera* (pelecypoda: Cypricardiacea) from the Upper Triassic of New Zealand. NZ J Geol Geophys 3:74–80

Grant-Mackie JA (1976a) The Upper Triassic bivalve *Monotis* in the southwest Pacific. Pac Geol 11:47–56

Grant-Mackie JA (1976b) Upper Jurassic fossils from the Waipapa Group of Tawharanui Peninsula, north Auckland, New Zealand. NZ J Geol Geophys 19:21–34

Grant-Mackie JA (1978a) Subgenera of the Upper Triassic bivalve *Monotis*. NZ J Geol Geophys 21:97–111

Grant-Mackie JA (1978b) Status and identity of the New Zealand Upper Triassic bivalve *Monotis salinaria* var. *richmondiana* Zittel 1866. NZ J Geol Geophys 21:375–402

Grant-Mackie JA (1978c) Systematics of New Zealand *Monotis* (Upper Triassic Bivalvia)— Subgenus *Entomonotis*. NZ J Geol Geophys 21:483–502

Grant-Mackie JA (1978d) Systematics of New Zealand *Monotis* (Upper Triassic Bivalvia): Subgenus *Maorimonotis*. J Roy Soc NZ 8:293–322

Grant-Mackie JA (1980a) Systematics of New Zealand *Monotis* (Upper Triassic Bivalvia)— Subgenus *Inflatomonotis*. NZ J Geol Geophys 23:629–637

Grant-Mackie JA (1980b) Systematics of New Zealand *Monotis* (Upper Triassic Bivalvia)— Subgenus *Eomonotis*. NZ J Geol Geophys 23:639–663

Grant-Mackie JA (1994) Mesozoic Bivalvia from Clerke and Mermaid Canyons, northwest Australian continental slope. J Austral Geol Geophys 15:119–125

Grant-Mackie JA (2011) A new early Jurassic *Otapiria* species (Monotidae: Bivalvia) from Murihiku rocks of the North Island of New Zealand. NZ J Geol Geophys 54:53–67

Grant-Mackie JA, Silberling NJ (1990) New data on the Upper Triassic bivalve *Monotis* in North America, and the new subgenus *Pacimonotis*. J Paleont 64:240–254

Grant-Mackie JA, Aita Y, Balme BE, Campbell HJ, Challinor AB, MacFarlan DAB, Molnar RE, Stevens GR, Thulborn RA (2000) Jurassic palaeobiogeography of Australasia. In: Wright AJ, Young GC, Talent JA, Laurie JR (eds) Palaeobiogeography of Australasian faunas and floras. Mem Assoc Australas Palaeontol 23:311–353

Guzmán G (1984) Los grifeidos infracretácicos *Aetostreon couloni* y *Ceratostreon boussingaulti*, de la Formación Rosablanca, como indicadores de oscilaciones marinas. Publ Geol, Ingeominas (Colombia):1–16

Hallam A (1969) Faunal realms and facies in the Jurassic. Palaeontology 12:1–18

Hallam A (1971) Provinciality in Jurassic faunas in relation to facies and palaeogeography. In: Middlemiss FA, Rawson PF, Newall G (eds) Faunal provinces in space and time. Geol J Spec Issue 4:129–152

Hallam A (1977) Jurassic bivalve biogeography. Paleobiology 3:58–73

Hallam A (1983) Early and mid-Jurassic molluscan biogeography and the establishment of the central Atlantic seaway. Palaeogeogr Palaeoclimatol Palaeoecol 43:181–193

Hallam A (1996) Recovery of the marine fauna in Europe after the end-Triassic and early Toarcian mass extinctions. In: Hart MB (ed) Biotic recovery from mass extinction events. Geol Soc Spec Publ 102:231–236

Hallam A, Miller AI (1988) Extinction and survival in the Bivalvia. In: Larwood GP (ed) Extinction and survival in the fossil record. Syst Assoc Spec 34:121–138

Hallam A, Biró-Bagóczky L, Pérez E (1986) Facies analysis of the Lo Valdés Formation (Tithonian-Hauterivian) of the High Cordillera of central Chile, and the palaeogeographic evolution of the Andean Basin. Geol Mag 123:425–435

Harrington HJ (1961) Geology of parts of Antofagasta and Atacama Provinces, northern Chile. Bull Am Assoc Petrol Geol 45:169–197

Haupt O (1907) Beiträge zur Fauna des oberen Malm und der unteren Kreide in der argentinischen Kordillere. N Jb Geol Paläontol BB 23:187–236

Hautmann M (2001a) Die Muschelfauna der Nayband-Formation (Obertrias, Nor-Rhät) des östlichen Zentraliran. Beringeria 29:3–181

Hautmann M (2001b) Taxonomy and phylogeny of cementing Triassic bivalves (Families Prospondylidae, Plicatulidae, Dimyidae and Ostreidae). Palaeontology 44:339–373

Hayami I (1984) Jurassic Marine Bivalve Faunas and Biogeography in Southeast Asia. Geol Palaeontol SE Asia 25:229–237

Hayami I (1987) Geohistorical background of Wallace's line and Jurassic Marine biogeography. In: Taira A, Tashiro M (eds) Historical biogeography and plate Tectonic evolution of Japan and Eastern Asia, Tokyo

Hayami I (1989) Outlook of the post-Paleozoic historical biogeography of pectinids in the Western Pacific Region. The Univ Mus, Univ Tokyo. Nature Culture 1:3–25

Hayami I, Maeda S, Ruiz Fuller C (1977) Some late Triassic Bivalvia and Gastropoda from the Domeyko range of north Chile. Trans Proc Palaeontol Soc Jpn NS 108:202–221

Hikuroa DCH (2005) The fauna and biostratigraphy of the Jurassic Latady Formation, Antarctic Peninsula. Univ Auckland Ph D Thesis (unpublished)

Hikuroa D, Grant-Mackie JA (2008) New species of late Jurassic Australobuchia (Bivalvia) from the Murihiku Terrane, Port Waikato-Kawhia region, New Zealand. Alcheringa 32:73–98

Hillebrandt A (1981) Kontinentalverschiebung und die paläozoogeographischen Beziehungen des südamerikanischen Lias. Geol Runds 70:570–582

Hillebrandt A, Westermann GEG, Callomon JH, Detterman RL (1992) Ammonites of the circum-Pacific region. In: Westermann GEG (ed) The Jurassic of the circum-Pacific. Cambridge University Press, New York

Howarth MK, Morris NJ (1998) The Jurassic and Lower Cretaceous of Wadi Hajar, Southern Yemen. Bull Nat Hist Mus (Geol) 54(1):1–32

Hudson N (1999) The middle Jurassic of New Zealand. A study of the Lithostratigraphy and Biostratigraphy of the Ururoan, Temaikan and Lower Heterian Stages (?Pliensbachian to ?Kimmeridgian). Univ Auckland Ph D Thesis (unpublished)

Hudson RGS, Jefferies RPS (1961) Upper Triassic brachiopods and lamellibranchs from the Oman Peninsula, Arabia. Palaeontology 4:1–41

Jaboli D (1959) Fossili giurassici dell'Harar (Africa Orientale). Brachiopodi, Lamellibranchi e Gasteropodi. Accad Naz Lincei 4. Docum Paleontol 1:1–100

Jaitly AK (1986a) Revised morphotaxonomic description of six Middle Jurassic pholadomyoid clams from Kala Dongar, Pachchham Island, District Kachchh, Gujarat. Indian Min 40:39–46

Jaitly AK (1986b) Indomya, a new subgenus of Pholadomya from the Middle Jurassic of Kachchh, Western India (Bivalvia: Pholadomyidae). Veliger 28:457–459

Jaitly AK (1986c) Middle Jurassic limids from Kaladongar, Pachchham Island, Kachchh, Gujarat. Q J Geol Mineral Met Soc India 58:42–52

Jaitly AK (1988) Some Middle Jurassic clams from Kala Dongar, Pachchham Island, Kachchh, Gujarat. Indian Min 42:117–125

Jaitly AK (1989) Some rare veneroid clams from Middle Jurassic rocks of Kala Dongar, Pachchham Island, District Kachchh (Gujarat). Proc Indian Nat Sci Acad 55A:570–577

Jaitly AK (1992) Neocrassinid bivalves (Heterodonta) from the Middle Bathonian (Jurassic) of Kachchh, Western India. Paläontol Z 66:67–79

Jaitly AK, Singh CSP (1983) A new species of Pronoella Fischer (Bivalvia) from the Bathonian (Middle Jurassic) rocks of Kaladongar, Pachchham Island, Kachchh. J Geol Soc India 24:476–478

Jaitly AK, Fürsich FT, Heinze M (1995) Contributions to the Jurassic of Kachchh, western India. IV. The bivalve fauna. Part I. Subclasses Palaeotaxodonta, Pteriomorphia, and Isofilibranchia. Beringeria 16:147–257

Jaworski E (1914) Beiträge zur Kenntnis des Jura in Süd–Amerika. Teil I. Allgemeiner Teil. N Jb Mineral, Geol Paläontol BB 37:285–342

Jaworski E (1915) Beiträge zur Kenntnis des Jura in Süd–Amerika. Teil II. Spezieller, paläontologischer Teil. N Jb Mineral Geol Paläontol BB 40:364–456

Jaworski E (1922) Die marine Trias in Südamerika. In: Steinmann G (ed) Beiträge zur Geologie und Paläontologie von Südamerika. N Jb Mineral Geol Paläontol BB 47:93–200

Jeletzky JA (1983) Macroinvertebrate paleontology, biochronology, and paleoenvironments of Lower Cretaceous and upper Jurassic rocks, Deep Sea Drilling Hole 511, eastern Falkland Plateu. Initial Rep Deep Sea Drill Proj 71:951–975

Jones DL, Plafker J (1976) Mesozoic megafossils from DSDP Hole 327A and Site 330 on the Eastern Falkland Plateau. Init Rep Deep Sea Drill Proj 36:845–855

Jordan R (1971) Megafossilien des Jura aus dem Antalo-Kalk von Nord-Äthiopien. Geol Jb 116:141–171

Kalantari A (1981) Iranian fossils. Min Oil, Nat Iran Oil Co (Tehran). Geol Lab Publ 9:1–35

Kanjilal S (1979a) Jurassic Camptonectes (Bivalvia) from the Habo Hill, District Kutch (Gujarat, W. India). J Mollusc Stud 45:115–124

Kanjilal S (1979b) Studies on some less known bivalves from the Jurassic rocks of Habo Hill, Kutch. Bull Earth Sci 7:23–32

Kanjilal S (1980a) Notes on two new species of Bivalvia from the Jurassic rocks of Habo Hill in Kutch. J Geol Soc India 21:249–252

Kanjilal S (1980b) Studies on Jurassic nuculids (Bivalvia) from the Habo Hill, District Kutch, Gujarat. Proc 3rd Indian Geol Congr (Poona 1980):331–346

Kanjilal S (1981) On some pteriomorph and heterodont Bivalvia from the Jurassic rocks of Habo Hill, District Kutch (Gujarat), W India. Proc Indian Natl Sci Acad B 46:264–286

Kanjilal S, Singh CSP (1973) A new nuculanid genus from the Callovian of Kutch (Gujarat), India. Proc Malacol Soc London 40:469–471

Kanjilal S, Singh CSP (1980) Studies on the Jurassic oxytomids (Bivalvia) from the Habo Hill in Kutch, W. India. J Palaeontol Soc India 23–24:16–22

Kauffman EG (1973) Cretaceous Bivalvia. In: Hallam A (ed) Atlas of Palaeobiogeography. Elsevier, Amsterdam

Kelly SRA (1995) New Trigonioid bivalves from the Early Jurassic to the earliest Cretaceous of the Antarctic Peninsula region: systematics and Austral paleobiogeography. J Paleontol 69:66–84

Kiessling W, Scasso R (1996) Ecological perspectives of late Jurassic radiolarian faunas from the Antarctic Peninsula. In: Riccardi AC (ed) Advances in Jurassic Research. GeoResearch Forum 1–2:317–326

Kiessling W, Pandey DK, Schemm-Gregory M, Mewis H, Aberhan M (2011) Marine benthonic invertebrates from the Upper Jurassic of northern Ethiopia and their biogeographic affinities. J Afric Earth Sci 59:195–214

Kitchin FL (1903) The Jurassic fauna of Cutch, the Lamellibranchiata; No. 1, Genus Trigonia. Mem Geol Surv India, Palaeontol Indica 9, 3, 2:1–122

Kluyver HM, Tirrul R, Chance PN, Johns GW, Meixner HM (1978) Explanatory text of the Naybandan Quadrangle Map 1:250,000. 1–143

Körner K (1937) Marine (Cassianer-Raibler) Trias am Nevado de Acrotambo (Nord-Peru). Palaeontographica A 86:145–237

Kristan-Tollmann E, Tollmann A, Hamedani A (1980) Beiträge zur Kenntnis der Trias von Persien. II. Zur Rhätfauna von Bagerabad bei Isfahan (Korallen, Ostracoden). Mitt österreich geol Gesells 73:163–235

Krystyn L, Richoz S, Baud A, Twitchett R (2003) A unique Permian-Triassic boundary section from the Neotethyan Hawasina Basin, Central Oman Mountains. Palaeogeogr Palaeoclimatol Palaeoecol 191:329–344

Lambert LR (1944) Algunas trigonias del Neuquén. Rev Mus La Plata (ns) Paleontol 2, 14:357–397

Leanza AF (1941) Apuntes estratigráficos sobre la región cruzada por el curso inferior del arroyo Carrín-Curá, en el Neuquén (Patagonia). Notas Mus La Plata Geol 6(13):203–213

Leanza AF (1968) Anotaciones sobre los fósiles jurásico-cretácicos de Patagonia Austral (colección Feruglio) conservados en la Universidad de Bologna. Acta Geol Lilloana 9:121–179

Leanza HA (1993) Jurassic and Cretaceous Trigoniid Bivalves from West-Central Argentina. Bull Am Paleontol 105(343):1–95

Leanza HA, Garate–Zubillaga JI (1987) Fauna de Trigonias (Bivalvia) del Jurásico y Cretácico inferior de la Provincia del Neuquén, Argentina, conservadas en el Museo Juan Olsacher de

Zapala. In: Volkheimer W (ed) Bioestratigrafía de los sistemas regionales del Jurásico y Cretácico de América del Sur, 1. Mendoza: 201–255

Levy R (1967) Revisión de las Trigonias de Argentina. Parte IV. Los Megatrigoniinae de Argentina y su relación con *Anditrigonia* gen. nov. Ameghiniana 5:135–144

Li X (1990) The Marine Jurassic and Lower Cretaceous of Southern Xizang (Tibet): Bivalve Assemblages, Correlation, Paleoenvironments and Paleogeography. Univ Auckland Ph D Thesis (unpublished)

Li X, Grant-Mackie JA (1988) Upper Jurassic and Lower Cretaceous *Buchia* (Bivalvia) from Southern Tibet, and some wider considerations. Alcheringa 12:249–268

Li X, Grant-Mackie JA (1994) New Middle Jurassic-Lower Cretaceous bivalves from southern Tibet. J SE Asian Earth Sci 9(3):263–276

Lieberman BS (2003) Paleobiogeography: the relevance of fossils to biogeography. An Rev Ecol Syst 34:51–69

Liu C (1995) Jurassic bivalve palaeobiogeography of the Proto-Atlantic and the application of multivariate analysis methods in palaeobiogeography. Beringeria 16:3–123

Liu C, Heinze M, Fürsich FT (1998) Bivalve provinces in the Proto-Atlantic and along the southern margin of the Tethys in the Jurassic. Palaeogeogr Palaeoclimatol Palaeoecol 137:127–151

Lo Forte GL (1988) La fauna de trigonias (Mollusca; Bivalvia) del Tithoniano-Neocomiano de la Quebrada Blanca, Alta Cordillera de Mendoza. Actas 5to. Congr Geol Chileno (Santiago) 2:C277–C293

Macchioni F, Cecca F (2002) Biodiversity and biogeography of middle-late liassic ammonoids: implications for the early Toarcian mass extinction. Géobios Mém Spec 24:165–175

MacFarlan DAB (1992) Triassic & Jurassic Rhynchonellacea (Brachiopoda) from New Zealand & New Caledonia. Roy Soc NZ Bull 31:1–310

Makridin VP (1973) The principles of discrimination and nomenclature of subdivisions in the paleogeographic zoning of marine basins. Paleontol J 7:27–131

Malchus N, Aberhan A (1998) Transitional gryphaeate/exogyrate oysters (Bivalvia: Gryphaeidae) from the Lower Jurassic of northern Chile. J Paleontol 72:619–631

Manceñido MO (2002) Paleobiogeography of Mesozoic brachiopod faunas from Andean-Patagonian areas in a global context. Géobios MS 24:176–192

Manceñido MO, Dagys AS (1992) Brachiopods of the circum-Pacific region. In: Westermann GEG (ed) The Jurassic of the circum-Pacific. Cambridge University Press, New York

Manivit J, Le Nindre YM, Vaslet D (1990) Le Jurassique d'Arabie Centrale. vol 4. Histoire géologique de la bordure occidentale de la plate-forme arabe. Doc BRGM 194:1–559

Marwick J (1935) Some new genera of the Myalinidae and Pteridae of New Zealand. Trans Proc Roy Soc NZ 65:295–303

Marwick J (1953) Divisions and faunas of the Hokonui system (Triassic and Jurassic). NZ Geol Surv Palaeontol Bull 21:1–142

Marwick J (1956) Three Fossil Mollusca from the Hokonui System (Triassic and Jurassic). Trans Roy Soc NZ 83:489–491

Masse JP (1992) The Lower Cretaceous Mesogean benthic ecosystems: palaeoecologic aspects and palaeobiogeographic implications. Palaeogeogr Palaeoclimatol Palaeoecol 91:331–345

Medina FA, Ramos AM (1983) Geología de las inmediaciones del refugio Ameghino (64° 26' S, 58° 59' W), Tierra de San Martín, Península Antártica. Inst Antárt Argent Contrib 229:1–14

Mette W (2004) Middle to Upper Jurassic sedimentary sequences and marine biota of the early Indian Ocean (Southwest Madagascar): some biostratigraphic, palaeoecologic and palaeobiogeographic conclusions. J Afric Earth Sci 38:331–342

Newton RB (1889) Notes on fossils from Madagascar, with descriptions of two new species of Jurassic Pelecypoda from that island. Q J Geol Soc Lond 45:331–338

Newton RB (1895) On a collection of fossils from Madagascar obtained by Rev. R. Baron. Q J Geol Soc London 51:72–91

Nicola M (1950–1951) Paléontologie de Madagascar. XXIX. Étude de quelques gisements fossilifères du sud-ouest de Madagascar. Ann Paléontol 36:141–168; 37:1–46

Niu Y, Jiang B, Huang H (2011) Triassic marine biogeography constrains the palaeogeographic reconstruction of Tibet and adjacent areas. Palaeogeogr Palaeoclimatol Palaeoecol 306:160–175

Olivero EB (1988) Cefalópodos y bivalvos titonianos y hauterivianos de la Formación Lago La Plata, Chubut. Ameghiniana 24:181–202

Pandey DK, Agrawal SK (1984) Bathonian-Callovian molluscs of Gora Dongar, Pachchham 'Island'. Q J Geol Min Metal Soc India 56:176–197

Pandey DK, Fürsich FT, Heinze M (1996) Contributions to the Jurassic of Kachchh, Western India. V. The bivalve fauna. Part II. Subclass Anomalodesmata. Beringeria 18:51–87

Pérez E, Reyes R (1977) Las trigonias jurásicas de Chile y su valor cronoestratigráfico. Bol Inst Investig Geol Chile 30:1–58

Pérez E, Reyes R (1983) Las especies del género Anditrigonia Levy, 1967, en la colección Philippi. Rev Geol Chile 18:15–41

Pérez E, Reyes R (1985) Presencia de Linotrigonia van Hoepen (Bivalvia; Trigoniidae) en el Kimmeridgiano del norte de Chile. Rev Geol Chile 25–26:135–143

Pérez E, Reyes R (1986) Presencia de Buchotrigonia (Syrotrigonia) Cox, 1952 (Bivalvia; Trigoniidae) en Sudamérica y descripción de dos especies nuevas. Rev Geol Chile 28–29:77–93

Pérez E, Reyes R (1991) El Orden Trigonioida (Mollusca; Bivalvia) en el Mesozoico de Sudamérica. Res, 6 to. Congr Geol Chileno: 72–76

Pérez E, Biro L, Reyes R (1987) Nuevos antecedentes sobre Virgotrigonia Alleman, 1985 (Bivalvia; Trigoniidae) y presencia de V. hugoi (Leanza) en Chile. Rev Geol Chile 30:35–45

Pérez E, Aberhan M, Reyes R, Hillebrandt A (2008) Early Jurassic Bivalvia of northern Chile. Part III. Order Trigonioida. Beringeria 39:51–102

Prinz P (1985) Stratigraphie und Ammonitenfauna der Pucara-Gruppe (Obertrias-Unterjura) von Nord-Peru. Palaeontographica A 188:153–197

Quilty PG (1978) Late Jurassic bivalves from Ellsworth Land, Antarctica: their systematics and paleogeographic implications. NZ J Geol Geophys 20:1033–1080

Quilty PG (1982) Tectonic and other implications of middle–upper Jurassic Rocks and Marine Faunas from Ellsworth Land, Antarctica. In: Craddock C (ed) Antarctic geoscience. The University of Wisconsin Press, Madison

Quilty PG (1983) Bajocian bivalves from Ellsworth Land, Antarctica. NZ J Geol Geophys 26:395–418

Riccardi AC (1977) Berriasian invertebrate fauna from the Springhill Formation of Southern Patagonia. N Jb Geol Paläontol Abh 155:216–252

Riccardi AC (1988) The Cretaceous system of southern South America. Geol Soc Ame Mem 168:1–161

Riccardi AC (1991) Jurassic and Cretaceous marine connections between the Southeast Pacific and Tethys. Palaeogeogr Palaeoclimatol Palaeoecol 87:155–189

Riccardi AC, Damborenea SE, Manceñido MO (1990a) Lower Jurassic of South America and Antarctic Peninsula. In: Westermann GEG, Riccardi AC (eds) Jurassic taxa ranges and correlation charts for the Circum Pacific. 3. South America and Antarctic Peninsula. Newsl Stratigr 21:75–103

Riccardi AC, Westermann GEG, Damborenea SE (1990b) Middle Jurassic of South America and Antarctic Peninsula. In: Westermann GEG, Riccardi AC (eds) Jurassic taxa ranges and correlation charts for the Circum Pacific. 3. South America and Antarctic Peninsu-la. Newsl Stratigr 21:105–128

Riccardi AC, Leanza HA, Volkheimer W (1990c) Upper Jurassic of South America and Antarctic Peninsula. In: Westermann GEG, Riccardi AC (eds) Jurassic taxa ranges and correlation charts for the Circum Pacific. 3. South America and Antarctic Peninsu-la. Newsl Stratigr 21:129–147

Riccardi AC, Damborenea SE, Manceñido MO, Scasso R, Lanés S, Iglesia Llanos MP (1997) Primer registro de Triásico marino fosilífero de la Argentina. Rev Asoc Geol Argent 52:228–234

Riccardi AC, Damborenea SE, Manceñido MO, Iglesia-Llanos MP (2004) The Triassic/Jurassic boundary in the Andes of Argentina. Riv Ital Paleontol Stratigra110(1):69–76

Riccardi AC, Damborenea SE, Manceñido MO, Leanza HA (2011) Megainvertebrados del Jurásico y su importancia geobiológica. In: Leanza HA, Arregui C, Carbone O, Danieli JC, Vallés JM (eds) Geología y Recursos Naturales de la Provincia del Neuquén, Relat 18° Congr Geol Argent: 441–464

Riley TR, Crame JA, Thomson MRA, Cantrill DJ (1997) Late Jurassic (Kimmeridgian–Tithonian) macrofossil assemblage from Jason Peninsula, Graham Land: evidence for a significant northward extension of the Latady Formation. Antarct Sci 9:434–442

Romero L, Aldana M, Rangel C, Villavicencio E, Ramírez J (1995) Fauna y flora fósil del Perú. Bol Inst Geol Min Metal, Ser D: Est Espec 17:1–332

Ros S, Echevarría J (2011) Bivalves and evolutionary resilience: old skills and new strategies to recover from the P/T and T/J extinction events. Histor Biol 23:411–429

Rossi Ronchetti C (1970) New contributions to the knowledge of the Jurassic fauna of Karkar (Northeast Afghanistan). Ital Exped, Karakorum (K2) Hindu Kush. Sci Rep IV, Paleontol–Zool Bot 2:43–74

Rossi-Ronchetti C, Fantini-Sestini N (1961) La fauna giurassica di Karkar (Afghanistan). Riv Ital Paleontol Stratigr 62(2):103–152

Rubilar A (1998) La Superfamilia Ostreacea en Chile y su importancia cronoestratigráfica, paleogeográfica y paleoecológica (Triásico superior-Jurásico). Ph D Thesis, Univ Nac La Plata (Unpublished)

Sato T, Westermann GEG, Skwarko SK, Hasibuan F (1978) Jurassic biostratigraphy of the Sula Islands, Indonesia. Bull Geol Surv Indonesia 4:1–28

Schairer G, Seyed-Emami K, Fürsich FT, Senowbari-Daryan B, Aghanabati SA, Majidifard MR (2000) Stratigraphy, facies analysis and ammonite fauna of the Qal'eh Dokhtar Formation (Middle-Upper Jurassic) at the type locality west of Boshrouyeh (east-central Iran). N Jb Geol Paläontol Abh 216:35–66

Shi GR, Grunt TA (2000) Permian Gondwana-Boreal antitropicality with special reference to brachiopod faunas. Palaeogeogr Palaeoclimatol Palaeoecol 155:239–263

Singh CSP, Kanjilal S (1974) Some fossil mussels from the Jurassic rocks of Habo Hill in Kutch, Gujarat, western India. Indian J Earth Sc 1:113–125

Singh CSP, Kanjilal S (1977) *Habonucula*, a new Nuculid (Bivalvia) genus from Jurassic rocks of Kutch (Gujarat), W. India. J Geol Soc India 18:189–193

Singh CSP, Kanjilal S (1982) On some Jurassic astartid bivalves from the Habo Hill in Kutch. J Palaeontol Soc India 27:49–61

Singh CSP, Rai JN (1980) Bathonian-Callovian fauna of Western Bela Island (Kutch). Part I. Bivalve families Cardiidae, Neomiodontidae, Corbulidae. J Palaeontol Soc India 23–24:71–80

Singh CSP, Jaitly AK, Pandey DK (1982) A new middle Jurassic Bivalve Genus, *Agrawalimya*, from Kachchh (Gujarat), India. Veliger 24:273–275

Skwarko SK (1967) Mesozoic Mollusca from Australia and New Guinea. Bull Bur Min Res Geol Geophys 75:1–101

Skwarko SK (1973) First report of Domerian (Lower Jurassic) marine Mollusca from New Guinea. Bull Bur Min Res Geol Geophys 140:105–112

Skwarko SK (1974) Jurassic fossils of Western Australia, 1: Bajocian Bivalvia of the Newmarracarra Limestone and the Kojarena Sandstone. Bull Bur Min Res Geol Geophys 150:1–53

Skwarko SK (1981a) A new upper Mesozoic trigoniid from western Papua New Guinea. Bull Bur Min Res Geol Geophys 209:53–55

Skwarko SK (1981b) *Spia*, a new Triassic Bakevellid bivalve from Papua New Guinea. Bull Bur Min Res Geol Geophys 209:63–64

Skwarko SK (1983) *Somareoides hastatus* (Skwarko), a new Late Triassic bivalve from Papua New Guinea. Bull Bur Min Res Geol Geophys 217:67–68

Skwarko SK, Nicholl R, Campbell KSW (1976) The Late Triassic molluscs, conodonts and brachiopods of the Kuta Formation, Papua New Guinea. J Austral Geol Geophys 1:219–230

Smith PL (1989) Paleobiogeography and Plate Tectonics. Geosci Canada 15:261–279

Sohl NF (1987) Cretaceous gastropods: contrasts between Tethys and the temperate provinces. J Paleontol 61:1085–1111

Sokolov DN (1946) Algunos fósiles suprajurásicos de la República Argentina. Rev Soc Geol Argent 1:7–16

Speden IG (1970) Three new inoceramid species from the Jurassic of New Zealand. NZ J Geol Geophys 13:825–851

Speden IG, Keyes IW (1981) Illustrations of New Zealand Fossils. New Zealand Department of Scientific and Industrial Research, DSIR Information Series 150, Wellington

Stefanini G (1939) Molluschi del Giuralias della Somalia, Gasteropodi e Lamellibranchi. Palaeontograph Ital 32(4):103–270

Stehn E (1923) Beiträge zur Kenntnis des Bathonien und Callovien in Südamerika. N Jb Min Geol Paläontol BB 48:52–158

Stevens GR (1963) Faunal realms in Jurassic and Cretaceous belemnites. Geol Mag 100:481–493

Stevens GR (1967) Upper Jurassic fossils from Ellsworth Land, West Antarctica, and notes on Upper Jurassic biogeography of the South Pacific region. NZ J Geol Geophys 10:345–393

Stevens GR (1973) Jurassic belemnites. In: Hallam A (ed) Atlas of palaeobiogeography. Elsevier, Amsterdam

Stevens GR (1977) Mesozoic biogeography of the South-West Pacific and its relationship to plate tectonics. Internatl Symp Geodynamics SW Pacific. Ed. Technip, Paris: 309–326

Stevens GR (1978) Jurassic. Paleontology. In: Suggate RP, Stevens GR, Te Punga MT (eds) The geology of New Zealand. NZ Geol Surv, 1:215–228

Stevens GR (1980) Southwest Pacific faunal palaeobiogeography in Mesozoic and Cenozoic times: a review. Palaeogeogr PalaeoclimatolPalaeoecol 31:153–196

Stevens GR (1989) The nature and timing of biotic links between New Zealand and Antarctica in Mesozoic and early Cenozoic times. In: Crame JA (ed) Origins and evolution of the Antarctic biota. Geol Soc Spec Publ 47:141–166

Stevens GR (1990) The influences of palaeogeography, tectonism and eustasy on faunal development in the Jurassic of New Zealand. Atti Sec Conv Internaz Fossil Evol Ambiente (Pergola 1987):441–457

Teichert C (1940) Marine Jurassic in the North–West Basin, Western Australia. J Roy Soc W Austral 26:17–27

Thevenin A (1908) Paléontologie de Madagascar. V. Fossiles Liassiques. Ann Paléontol 3:105–143

Thiele Cartagena R (1967) El Triásico y Jurásico del Departamento de Curepto en la provincia de Talca. Publ Univ Chile Fac Cienc Fís Matem Dep Geol 28:27–46

Thomson MRA (1975a) Fossils from the South Orkney Islands: I. Coronation Island. Bull Brit Antarct Surv 40:15–21

Thomson MRA (1975b) Upper Jurassic Mollusca from Carse Point, Palmer Land. Bull Brit Antarct Surv 41–42:31–42

Thomson MRA (1981) Late Mesozoic stratigraphy and invertebrate palaeontology of the South Orkney Islands. Bull Brit Antarct Surv 54:65–83

Thomson MRA (1982) Late Jurassic fossils from low Island, South Shetland Islands. Bull Brit Antarct Surv 56:25–35

Thomson MRA, Damborenea SE (1993) A new Middle Jurassic fauna from Antarctica. Arkell Internat Symp Jurassic Geol, Abstracts

Thomson MRA, Tranter TH (1986) Early Jurassic fossils from central Alexander Island and their geological setting. Bull Brit Antarct Surv 70:23–39

Thomson MRA, Willey LE (1972) Upper Jurassic and Lower Cretaceous *Inoceramus* (Bivalvia) from south–east Alexander Island. Bull Brit Antarct Surv 29:1–19

Tong J, Erwin DH (2001) Triassic Gastropods of the Southern Qinling Mountains, China. Smithson Contrib Paleobiol 92:1–47

Trechmann CT (1918) The Trias of New Zealand. Q J Geol Soc London 73:165–246

Trechmann CT (1923) The Jurassic rocks of New Zealand, with an Appendix on Ammonites from New Zealand by L.F. Spath. Q J Geol Soc London 79:246–312

Venzo S (1949) Il Batoniano a Trigonia dell'Oltregiuba settentrionale e del Borana sud-orientale (Africa Orientale) con osservazioni stratigrafiche sulla regione. Palaeontograph Ital 45:111–177

Waterhouse JB (2008) Aspects of the evolutionary record for fossils of the Bivalve subclass Pteriomorphia Beurlen. Earthwise 8:1–220

Weir J (1930) Mesozoic Brachiopoda and Mollusca from Mombasa. Monogr Geol Dep Hunterian Mus Glasgow Univ 4:73–102

Westermann GEG (1981) Ammonite biochronology and biogeography of the Circum Pacific Middle Jurassic. In: House MR, Senior JR (eds) The Ammonoidea. Syst Assoc Spec 18

Westermann GEG (1996) New Mid-Jurassic Ammonitina from New Zealand: implications for biogeography and oceanography. In: Riccardi AC (ed) Advances in Jurassic Research. GeoResearch Forum 1–2:179–185

Westermann GEG (2000a) Biochore classification and nomenclature in paleobiogeography: an attempt at order. Palaeogeogr Palaeoclimatol Palaeoecol 158:1–13

Westermann GEG (2000b) The marine faunal realms of the Mesozoic: review and revision under the new guidelines for biogeographic classification and nomenclature. Palaeogeogr Palaeoclimatol Palaeoecol 163:49–68

Westermann GEG, Hudson N (1991) The first find of Eurycephalitinae (Jurassic Ammonitina) in New Zealand and its biogeographic implications. J Paleontol 65:689–693

Westermann GEG, Riccardi AC, Palacios O, Rangel C (1980) Jurásico Medio en el Perú. Bol Inst Geol Min Metal Perú D 9:1–47

Whitehouse FW (1924) Some Jurassic fossils from Western Australia. J Roy Soc W Austral 11:1–13

Willey LE (1975a) Upper Jurassic and lower Cretaceous Grammatodontinae (Bivalvia) from southern Alexander Island. Bull Brit Antarct Surv 41(42):1–22

Willey LE (1975b) Upper Jurassic and lowest Cretaceous Trigoniidae (Bivalvia) from south–eastern Alexander Island. Bull Brit Antarct Surv 41(42):77–85

Willey LE (1975c) Upper Jurassic and lower Cretaceous Pinnidae (Bivalvia) from southern Alexander Island. Bull Brit Antarct Surv 41(42):121–131

Yancey TE, Stanley GD Jr, Piller WE, Woods MA (2005) Biogeography of the Late Triassic wallowaconchid megalodontoid bivalves. Lethaia 38:351–365

Ziegler MA (2001) Late Permian to Holocene Paleofacies evolution of the Arabian plate and its Hydrocarbon occurrences. Geo Arabia 6:445–504

Chapter 6
Global Scale

Abstract Southern Hemisphere bivalves have provided arguments for the analysis of some interesting topics of global significance, such as bipolarity and the establishment of seaways. The records of Triassic and Jurassic bivalves with bipolar distribution are numerous, and show that bipolarity was a persistent phenomenon in marine environments for hundreds of million years. These examples are potentially very enlightening for the discussion about the causes of this global disjunct distribution pattern, for which an integrative approach is still pending. Bivalves were also significant for the early proposal of a marine connection between western Tethys and eastern Panthalassa in the Early Jurassic, now known as Hispanic Corridor. The evolution of similarity coefficients during time has proven to be a good tool for detecting changes in the relationships of bivalve faunas from western Tethys and eastern Panthalassa. Within regions from the same side of the passage, similarity was high at all times and increased slightly with time during the Early and Middle Jurassic. When comparing regions from either side of the Hispanic Corridor, a sudden increase in similarity beginning very early during the Early Jurassic from low values during Late Triassic times is evident, indicating that the first marine connection between them was probably established by Pliensbachian times. The shallow connection acted first as a screen, being an effective barrier for "neritic" species, while allowing the passage of benthonic littoral species. This is also related to the change in water current pattern. The observed southwards shift of the boundary zones between Tethyan and South Pacific along the western South American coast during the Early Jurassic is coincident with a symmetric northwards shift of the Tethyan/Boreal boundary in the Northern Hemisphere, suggesting a common cause, such as global climate change.

Though this book is not treating Triassic/Jurassic bivalve distribution at a worldwide scale, the already discussed arguments about the persistence of a South Pacific biochorema through time (Chap. 5) have undoubted global interest for

S. E. Damborenea et al., *Southern Hemisphere Palaeobiogeography of Triassic-Jurassic Marine Bivalves*, SpringerBriefs Seaways and Landbridges: Southern Hemisphere Biogeographic Connections Through Time, DOI: 10.1007/978-94-007-5098-2_6, © The Author(s) 2013

understanding the broad distribution patterns of marine biotas and their evolution through time.

Furthermore, there are a few extra topics of universal significance which will be addressed here, since Southern Hemisphere bivalves provided, and will certainly continue to offer, considerable insight for their analysis. This chapter will deal mainly with two worldwide issues: bipolarity and the establishment of major seaways, but water circulation patterns and the boundaries of the main biochoremas will also be shortly discussed in relation to the evidence provided by the distribution of marine bivalves.

6.1 Bipolarity

Disjunct distributions have long interested biogeographers, since the existence of taxonomically related taxa living in areas separated by barriers calls for a historic explanation of their development. Among disjunct distributional patterns, one of the first to be recognized is that of taxa with northern and southern high-latitude components separated by a distinct low-latitude gap, which was observed both in terrestrial and marine biotas. This was already discussed by Darwin (1859), and is generally known as bipolarity.

Bipolarity and antitropicality are relatively common but yet poorly known large-scale biogeographic phenomena. Their explanation has been, and still is, highly controversial, since they have been interpreted as the result of either dispersal events from a center of origin or vicariant disruption (Cecca 2009), rationales which are directly linked, respectively, to two bitterly antagonist schools: dispersalism and vicariance. An integrative approach to understand bipolarity is certainly needed (Donoghue 2011), and examples from the past such as those discussed here can be very enlightening.

But let us leave the explanation of the causes of this phenomenon for the moment, and concentrate on the observed facts first.

6.1.1 Triassic–Jurassic Bipolar Bivalves

The characterization of marine faunas into a tropical (peri-equatorial) and two antitropical (polar) belts (these last frequently containing bipolar taxa) was demonstrated beyond doubt for the Cretaceous–Cenozoic (Kauffman 1973; Zinsmeister 1982) and for the Permian–Triassic (Shi and Grunt 2000), in addition to being revealed for the Triassic–Jurassic on the basis of bivalve distribution (Crame 1986, 1987, 1992, 1993; Damborenea and Manceñido 1992; Damborenea 1993, 2002a, b, and here).

Late Mesozoic (Late Jurassic–Cretaceous) bipolar bivalve faunas were extensively discussed as a consequence of the development on the knowledge of faunas from Antarctica, New Zealand, and South America (Crame 1986, 1987, 1992 and references therein), which revealed the presence of several taxa previously thought to be restricted to the northern high-latitude regions, such as *Buchia, Arctotis, Anopaea, Retroceramus,* and *Aucellina.*

A detailed account of the distribution of Early Jurassic pectinaceans from South America revealed the presence of at least eight genera with bipolar distributions during that time: *Otapiria, Palmoxytoma, Arctotis, Asoella, Kolymonectes, Radulonectites, Agerchlamys,* and *Harpax* (Damborenea 1993) and these were soon included in more general discussions on the subject by Crame (1993, 1996a, b). Bipolarity appears to have been less evident during the Middle Jurassic, though *Retroceramus* had very similar species in high latitudes from both hemispheres (Damborenea 1990) from Bathonian to Tithonian times.

Triassic and Jurassic bipolar bivalve genera had very similar species on both high-latitude areas (Fig. 6.1), and sometimes antitropical populations are even regarded as conspecific. It is also interesting to note that known examples are increasing as thorough and open-minded systematic revisions are performed, and many taxa originally thought to be restricted to the polar or high-latitudes regions of one hemisphere have proven to have a bipolar distribution instead as research progresses. On the other hand, the absence of these taxa in low-latitude areas cannot be blamed on poor knowledge, since the faunas from those areas are the best known.

It should also be noted that the distribution pattern of a particular genus may change with time, bipolarity being only apparent during part of the whole genus stratigraphic range. Detailed knowledge of the occurrences may add interesting points to the debate, since not always these observed changes can be attributed solely to an insufficient fossil record. To the pectinaceans, some other groups are now known to have had a bipolar distribution during the time interval involved (Crame 1993, Table 1), and a few examples will be shortly mentioned below.

Aparimella Campbell was an Anisian to Carnian halobiid, known from New Zealand during its whole time range, but reported also from the early Carnian of Svalbard (Campbell 1994).

Asoella Tokuyama (Fig. 6.1b, l) spanned the Triassic–Jurassic boundary; it ranged from Anisian to Pliensbachian and had mostly a circum-Pacific distribution (see discussion and references in Damborenea and Manceñido 2012 and Ros et al. 2012), though it was certainly more abundant both in the northern (Japan) and southern (New Zealand, Argentina) areas during its whole time range. It was also reported from China. Species of this genus were byssally attached and there is some taphonomic evidence of a facultative pseudoplanktonic habit for some of them.

Maoritrigonia Fleming was a late Triassic (Carnian–Rhaetian) trigoniid first regarded as endemic of the Maorian region (New Zealand–New Caledonia), but later its geographic range was extended to South America (Chile) and SW Alaska (see details and references in Ros et al. 2012). The poorly known *Minetrigonia* Kobayashi and Katayama was another late Triassic trigoniid with records in high-latitude regions of both hemispheres: Japan, Siberia, Alaska, British Columbia,

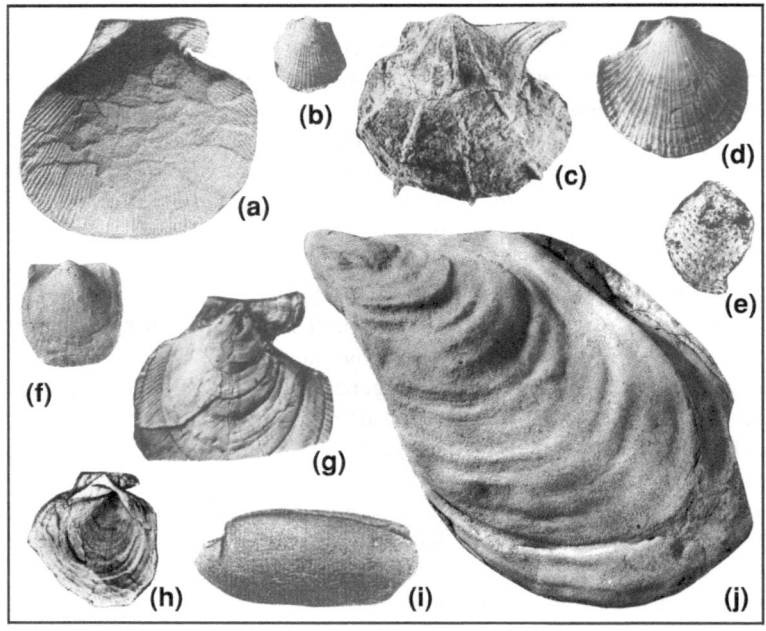

◄ **Fig. 6.1** Some examples of bipolar bivalves, **a–j**, specimens from northern high latitude regions. **a** *Agerchlamys propius* (Milova), Pliensbachian, NE Russia; **b** *Asoella confertoradiata* (Tokuyama), Carnian–Norian, Japan; **c** *Palmoxytoma cygnipes* (Phllips), Sinemurian, W Canada; **d** *Kolymonectes staeschei* (Polubotko), Sinemurian, NE Russia; **e** *Harpax spinosa* (Sowerby), Pliensbachian, NE Russia; **f** *Arctotis sublaevis* Bodyl., Bajocian, NE Russia; **g** *Radulonectites hayamii* Polubotko, Pliensbachian, NE Russia; **h** *Ochotochlamys bureiensis* Sey, Pliensbachian, NW Russia; **i** *Kalentera brodnayaensis* Milova, Pliensbachian, NE Russia; **j** *Retroceramus tongusensis* (Lahusen), upper Bajocian, NE Russia. **k–t**, Specimens from the southern high latitude regions. **k** *Agerchlamys wunschae* (Marwick), MLP23658, Pliensbachian, Argentina; **l** *Asoella campbellorum* Damborenea, MLP32730, Norian–Rhaetian, Argentina; **m** *Palmoxytoma* sp., OU17810, Aratauran, New Zealand; **n** *Kolymonectes weaveri* Damborenea, Pliensbachian, Argentina, MLP23686; **o** *Harpax rapa* (Bayle and Coquand), MLP16511, Pliensbachian, Argentina; **p** *Arctotis frenguellii* Damborenea, MLP10418, Early Jurassic, Argentina; **q** *Radulonectites sosneadoensis* (Weaver), MLP22321, Pliensbachian, Argentina; **r** *Ochotochlamys* sp., MLP27592, Pliensbachian, Argentina; **s** *Kalentera mackayi* Marwick, MLP 24873, Aratauran, New Zealand; **t** *Retroceramus stehni* Damborenea, MLP14672, Callovian, Argentina. **b** from Hayami 1975; **c** from Frebold 1957; **d, e, f, g, h, j** from Sey and Polubotko 1992; **i** from Milova 1988. Figures not to same scale, but proportional size differences are roughly shown

Chile, New Zealand, and probably Argentina (see Ros et al. 2012). A systematic revision may extend this distribution, since apparently some late Triassic species referred to *Trigonia* or *Myophoria* may actually belong to *Minetrigonia*.

Triaphorus Marwick was a late Triassic Kalenteridae originally described for New Zealand but later reported from New Caledonia, and also from Japan and NE Russia. It was probably a shallow burrower belonging to the shallow infauna or semi-infauna (see Ros et al. 2012). *Kalentera* Marwick (Fig. 6.1j, s) was also a shallow burrower Kalenteridae which lived from Norian to Toarcian times. Although initially thought to be restricted to Austral regions, later records show that it was endemic to the Maorian biochorema during the late Triassic, but had a bipolar distribution at high paleolatitudes during the early Jurassic (Damborenea 2004 and references therein), being known from NE Siberia, and reaching to northern Chile, but no species are known below 30° paleolatitude in either hemisphere.

Ochotomya Polubotko was a Norian–Rhaetian ceratomyid genus known both from northern (NE Siberia) and southern (New Zealand) regions. It was most probably a shallow burrower (see Ros et al. (2012) for references).

The early Jurassic genus *Palmoxytoma* Cox (Fig. 6.1c, m) was originally proposed for northern European species, but was later recorded from New Zealand and Argentina in austral regions; and from Canada, Japan, NE Siberia, and eastern Russia in boreal areas. It had a clear antitropical distribution during the Hettangian (New Zealand, Argentina and Chile in the South, NE Siberia in the North), while during Sinemurian and Pliensbachian times it seems to have been restricted to the Northern Hemisphere (Damborenea 1993 and references therein). Sporadic occurrences out of its high-latitude range have been used as indication of influx of cool seawater (see Arp and Seppelt 2012 for German Pliensbachian *Palmoxytoma*). The genus was reported from Europe (England, Sweden, France, Switzerland) and North America (Canada) but never from low paleolatitude regions.

Kolymonectes Milova and Polubotko (Fig. 6.1d, n) was traditionally mentioned as a typical element of Triassic-early Jurassic Boreal faunas, known from NE Russia and NW Canada. Moreover, it was used by some authors to define and outline the Arctic subprovince of the Boreal Realm. But at least during the late Pliensbachian the species *K. weaveri* Damborenea was very abundant in western Argentina and a related species was probably also present in New Zealand (Damborenea and Manceñido 1992; Damborenea 1993 and references therein).

Similarly, *Ochotochlamys* Milova and Polubotko (Fig. 6.1h, r) was originally thought to be endemic to the late Triassic–early Jurassic of NE Asia, but it was later found in Canada and the Pliensbachian deposits of southern Argentina (Damborenea 2002a and references therein), and thus a bipolar distribution for the late part of its range is proposed. Species of this genus were byssally attached.

Agerchlamys Damborenea (Fig. 6.1a, k) includes late Triassic–Early Jurassic species known from circum-Pacific localities. During most of this time range, but at least during the Pliensbachian, the genus had a clear bipolar distribution, with records from NE Russia, New Zealand, and west-central Argentina. Species of this genus are found in the Andes and in New Zealand in very fine-grained sediments, from marls to tuffaceous siltstones, but never in laminated dark mudstones. Although locally abundant in certain beds, they were not widespread, and at least *A. wunschae* (Marwick) seems to have been a stenotopic species limited to very low-energy but well-oxygenated environments. It was a byssally attached pectinid, and it was never associated to thick-shelled epifaunal bivalves or corals, but usually occurs with sponge spicules instead, probably sponges were their substrate (Damborenea 1993, 2002a and references therein).

Species of the genus *Harpax* Parkinson (or subgenus of *Plicatula* for some authors) were conspicuous elements of Boreal late Triassic and bipolar Early Jurassic (mainly Hettangian–Pliensbachian) bivalve faunas (Fig. 6.1e, o), reported from southern South America, New Zealand, NE Russia and Canada (Damborenea 2002a and references therein). This genus included species which cemented to hard substrates (mainly other shells) at least early in ontogeny, and adults could have been free recliners.

Radulonectites Hayami (Fig. 6.1g, q) was a byssally attached Early Jurassic pectinid genus. Since its original description for Japanese species, it was reported from the Hettangian–Sinemurian of the eastern Tethys, but during the Pliensbachian it was clearly bipolar (Argentina, East Siberia, Japan) (Damborenea 1993, 2002a and references therein).

Arctotis Bodylevsky (Fig. 6.1f, p) was also first thought to be typically boreal, known from Arctic regions from latest Early Jurassic to Late Jurassic times, but the discovery of late Jurassic *Arctotis* in Austral regions has changed this view, and a bipolar distribution for this genus during that time was proposed by Kelly (1984) and is now fully accepted with records from Antarctica (Crame 1993). The Toarcian/Aalenian species *Arctotis*? *frenguellii* Damborenea from western Argentina may indicate that the bipolar distribution extended in fact over the whole time range of the genus (Damborenea 1993 and references therein).

The bipolar occurrences of *Retroceramus* Koshelkina, *Anopaea* Eichwald and *Buchia* Rouillier (s.l.) were discussed by Crame (1993, for *Anopaea* see also Crame 1981 and Dhondt 1992).

6.1.2 Discussion of Triassic–Jurassic Bipolarity

The simplest explanation for such global pattern seems to be a latitudinally related climatic control, even if temperature gradients changed a lot with time and for the Jurassic they were milder than at present (thus generating wide transitional zones). This matter was extensively discussed by Crame (1986, 1987, 1992, 1993), Damborenea (1993, 2002a, b), and Sha (1996, 2003) with direct reference to Jurassic faunas, and little else can be added here. Sha (1996) suggested that this was the result of some degree of repeated interbreeding due to the regularity of the larval exchange and proposed both cold seawater temperatures and suitable substrate as the main environmental parameters which could control such past antitropical distributions.

It is also evident that this distribution pattern was not restricted to a particular mode of life; although many of the genera were probably byssally attached or free epifaunal recliners, some were cemented and some burrowers. Several could have been pseudoplanktonic and some facultative swimmers. They lived in a variety of normal marine environments, not being restricted to any particular substrate, energy, or water depth. Most of them probably had planktotrophic larval stages but actual data are still few.

On the other hand, they are not randomly distributed systematically; they do belong to a few families, notably Monotidae, Oxytomidae, Pectinidae, Kalenteridae, and Trigoniidae in the Triassic–Early Jurassic, with the addition of Buchiidae, Inoceramidae, and Retroceramidae in the Middle and Late Jurassic. Interestingly enough, these families are also those which have been already mentioned as generically more diverse than average at high latitudes (Fig. 5.2). Moreover, the Triassic–Jurassic bipolar bivalve set does not include a single genus of any of the families more diverse than average at low latitudes.

The records just discussed show that bipolarity was a persistent large-scale phenomenon in marine environments for hundreds of million years. Crame (1993) explained Jurassic and Cretaceous examples as the result of vicariance due to the disintegration of Pangea, but emphasized that phylogenetic studies of critical groups, as the Monotoidea, are needed to test this hypothesis. Such studies could also indicate whether convergent evolution occurred in unrelated stocks, but although this possibility cannot be altogether discarded, repetition of bipolarity makes this explanation hardly applicable to all known examples. In contrast, a mainly dispersalist explanation was favored by Sha (1996), who emphasized that planktotrophic larvae could use cool deep water currents for dispersal and even to maintain interbreeding between northern and southern populations in those examples where bipolarity is recognized at the specific level. Sha (2003) also argued that the pseudoplanktonic inferred habit for most of the bivalve genera involved was an efficient dispersal medium.

It is evident that there was a peak of bipolarity among bivalves during Early Jurassic times, which can be related to the disruption of the pre-existing water-movement patterns and the establishment of a circum-equatorial current which could have acted as a new barrier for the dispersal of some groups. Nevertheless, the repetition of the phenomenon through time both before and after this event makes it difficult to explain such global scale disjunct distributions just as a result of vicariance alone. Phylogenetic evidence of the groups involved and the Earth's tectonic history at these large spatial and time scales are still hard to match, and the question of the relentless repetition of bipolarity claims for an integrative and imaginative proposal. Crame (1993) recognized three main phases of bipolarity: Jurassic–Cretaceous, Paleogene–Early Neogene, and Plio-Pleistocene-Recent, and related each of them to different major causes, respectively, the disintegration of Pangea, global climatic change, and glacial cooling, with different emphasis on vicariance and dispersal.

It is evident that much more work is required in several fields of knowledge to advance in the integrated explanation of bipolarity. Data from the fossil record are essential, but a higher detail in both geographic and time ranges of the taxa involved is needed.

6.2 Seaways: The Hispanic Corridor

The opening of the North Atlantic, which triggered off the fragmentation of Pangea, and the establishment of a marine connection between the western Tethys and the eastern central Pacific (Hispanic Corridor) is one of the most important paleo-geographic events which occurred during the Jurassic. New biogeographic, geo-logic, paleomagnetic and geophysic data are constantly providing rich arguments for the discussion of this event. Among them, biogeography is a most powerful tool to understand its nature and timing, and was one of the first to be used in this connection.

The distribution patterns of bivalves during the late Triassic–Middle Jurassic time interval is now well known, both along the eastern Pacific (Damborenea 1996; Aberhan 1993, 1994, 1998a, b and references therein) and the western Tethys (Hallam 1977; Liu 1995; Liu et al. 1998 and references therein).

The distribution of certain bivalves was used to propose the hypothesis of the existence of a shallow marine connection (now called Hispanic Corridor) since Early Jurassic times (Damborenea and Manceñido 1979) and has been widely discussed since then on the basis of the distribution of bivalves and also other marine invertebrates (Hallam 1983; Smith and Tipper 1986; Nauss and Smith 1988; Newton 1988; Smith 1989; Smith et al. 1990; Riccardi 1991; Damborenea 2000; Aberhan 2001), especially in connection to the age of the establishment of a continuous marine corridor.

Fig. 6.2 Location of the
regions used for the analysis
of Hispanic Corridor. Base
map as in Fig. 1.2. 1,
Colombia; 2, Perú and N
Chile; 3, Central Chile and
Argentina; 4, Mexico; 5,
Cratonic USA and Sonomia
terrane; 6, North American
Boreal craton; 7, S France; 8,
Spain and Portugal; 9, Morocco and Algeria; 10, Japan;
11,Wallowa; 12, Wrangellia;
13, Stikinia; 14, Quesnellia;
15, Cadwallader

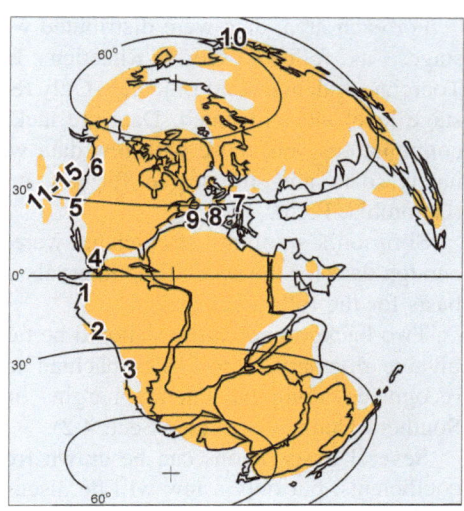

Some authors even proposed that the Hispanic Corridor might have been open
already in the Late Triassic, but supporting proof is doubtful and also there is strong
evidence against this idea, such as the paleobiogeographic distribution of the aberrant
bivalves included in the genus *Wallowaconcha*, known only from eastern Tethys and
eastern Panthalassa, with no occurrences in western Tethys (Yancey et al. 2005).

The quality of the database is critical in biogeographic approaches such as this.
A large amount of the data used here was compiled from papers published during
more than a century, and is necessarily uneven. Most published information was
re-evaluated, and for this reason the data were treated only at generic/subgeneric
level. This admittedly results in loss of information but is common practice in other
recent paleobiogeographic papers using bivalve data (see Liu 1995; Liu et al. 1998).

The distribution of 236 bivalve genera and subgenera was recorded for 15 geographic areas (Fig. 6.2), stage by stage from the Norian/Rhaetian to the Bajocian,
from the sources listed in Damborenea (2000) with the additions mentioned below.
Since the aim of the analysis was to register changes across the Hispanic Corridor,
some of the areas were pooled together: *South America*: (1 + 2 + 3), references
already mentioned in Sect. 5.1, and unpublished data; *Mexico*(4) can be pooled
together with cratonic North America, but it was not because this is a key area for the
discussion, data from Jaworski 1929, Sandoval and Westermann 1986, Stanley et al.
1994, Damborenea and González-León 1997, McRoberts 1997, Stanley and González-León 1997, and Scholz et al. 2008; *"Cratonic" North America*: (5 + 6);
Southwestern Tethys: (7 + 8 + 9). Apart from these regions, directly concerned
with the question of the Hispanic Corridor, data from Japan (10) were analyzed to
test the behavior of bivalve distribution across the Pacific. Although not directly
involved in the main subject, data from several suspect terranes of the western North
American margin were likewise included (11–15).

For each area, data were distributed within seven time slices corresponding to stages, as follows: Norian/Rhaetian, Hettangian, Sinemurian, Pliensbachian, Toarcian, Aalenian, and Bajocian. Only records with reliable age assignment to the stage level were included. Data are lacking or very scanty for some areas/time combinations, and those with few data were discarded for the analysis. Unfortunately, data are still scarce for key areas, such as northern South America (Colombia, Perú).

Simpson's similarity coefficients were calculated for every pair of areas with enough data for the seven time intervals considered. This set of coefficients is the basis for the following analysis.

Two background patterns should be taken into account: the overall increase in bivalve diversity at the Pliensbachian and paleolatitudinal control, which was recognized along the Pacific margins both for the northern (Tozer 1982) and Southern Hemispheres (see Sect. 4.2).

Several observations can be drawn from the analysis of this set of similarity coefficients, but only a few will be discussed here.

When analyzing the composition of the fauna between localities, there seems to be a general trend to increase the similarity value through time when localities situated at both extremes of the Corridor are compared (Fig. 6.3b), while between localities from the same side of the corridor values trend to be high but less variable (Fig. 6.3a). In order to test the significance of these observations, a GLM was performed. Since Simpson's coefficient of similarity is nothing but a proportion, and we are looking for general trends in the values of those proportions through time, a GLM is perfectly applicable. As a result, many of the localities (although not all of them) at both ends of the Corridor showed an increasing significant trend in similarity; some of these significant trends are shown in Fig. 6.3b.

On the other hand, among the comparisons between localities at the same extreme of the Corridor only a few showed some significant change through time, standing out Central Argentina and Chile, which tends to significantly increase the similarity with the Pacific localities of the Northern Hemisphere (Fig. 6.3a). Considering the expansion of tropical faunas during the Early Jurassic (Sect. 6.4), and the position of this locality close to the boundary between biogeographic units, this result was to some extent expected. But most important, this result shows that there might be many factors influencing the similarity between localities, and hence the observed patterns, though providing auxiliary support for the development of the Corridor, cannot be regarded as conclusive evidence on their own.

Across the Hispanic Corridor (Fig. 6.3b) there was a general pattern of sudden increase in similarity beginning very early during the Early Jurassic (Sinemurian) from low values during Late Triassic times, reaching similarity peaks at about Pliensbachian–Toarcian times, followed by a decrease by the end of the Early Jurassic. The maximum values then observed are similar to those obtained *within* regions and along the eastern Pacific for earlier times. Maximum similarity peaks across the Corridor fluctuate between the Pliensbachian and Toarcian for different regions.

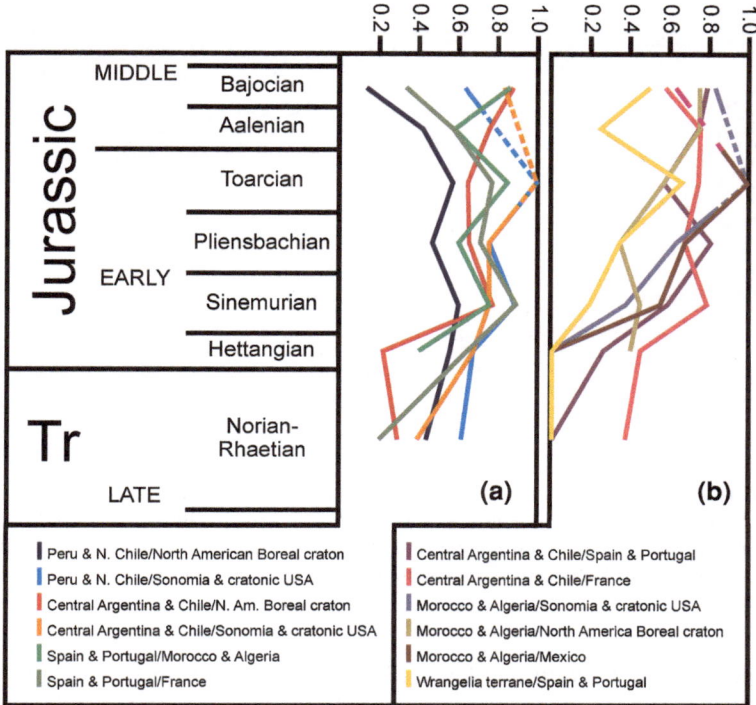

Fig. 6.3 Evolution of Simpson's similarity coefficients with time when the areas located in Fig. 6.2 are compared to each other. **a** Comparison of areas located within the same side of the Hispanic Corridor. **b** Comparison of areas situated on both sides of the Hispanic Corridor; *dashed lines* represent time intervals with 1 genus or none in one of the compared areas

The first marine connection was probably well established by Pliensbachian-Toarcian times, and thus preceded rifting by millions of years, but allowed an important faunal interchange of benthonic organisms such as bivalves. It is also evident that the shallow connection first acted as a screen, being an effective barrier for "neritic" species, while allowing the passage of benthonic littoral species (Damborenea and Manceñido 1979; Damborenea 2000; Aberhan 2001). It was postulated that the corridor might have been open much earlier, since early Hettangian times (Sha 2002), or even as early as Norian (Sandy and Stanley 1993, based on the distribution of brachiopods), but these last proposals are not conclusive, and in fact there is evidence against them (see discussion in Sect. 6.3).

Of special interest for this subject are taxa restricted to the western Tethys and the eastern Pacific and their mutual relationships. Several benthonic bivalve taxa, and particularly epifaunal stocks, can be followed "step by step" along this migration route, either eastwards or westwards, examples being *Weyla* and some other pectinoid species (Damborenea and Manceñido 1979, 1988; Aberhan 2001),

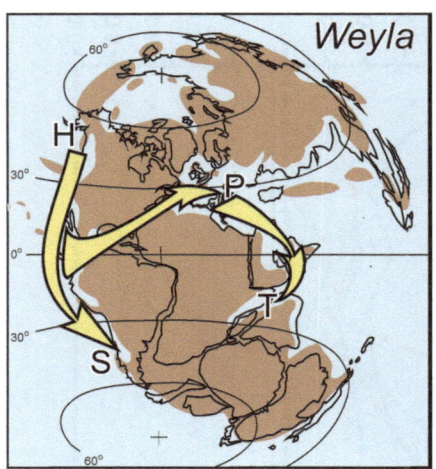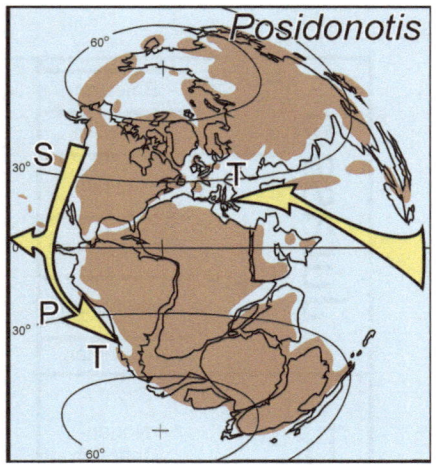

Fig. 6.4 Hypothetic dispersal routes proposed for early Jurassic bivalves belonging to different life-habit groups, exemplified with the sublittoral benthonic recliner *Weyla* (*left*) and the swimmer or pseudoplanktonic *Posidonotis* (*right*). First occurrences in the different regions are marked with letters: H, Hettangian, S, Sinemurian, P, Pliensbachian and T, Toarcian. Base map as in Fig. 1.2

and species of the *Lithiotis* reef fauna (Broglio Loriga and Neri 1976; Hillebrandt 1981; Nauss and Smith 1988; Aberhan and Hillebrandt 1999). *Weyla* migrated eastwards (Fig. 6.4) but other pectinoids and species of the *Lithiotis-Opisoma* association probably followed the same route at approximately the same time but in opposite direction. Other bivalve taxa which were not restricted to these regions but may have migrated along this corridor, according to the known distribution with time, are: from Tethys to eastern Pacific: *Palaeolopha, Pseudopecten, Atreta, Terquemia, Cardinia, Pteromya, Goniomya*; from eastern Pacific to Tethys *Gryphaea, Actinostreon, Preaexogyra, Aguilerella, Gervillaria, Lycettia*, and some trigoniids, such as *Frenguelliella, Jaworskiella, Vaugonia*, and *Psilotrigonia*.

Other possible consequences of the opening of the Hispanic Corridor for bivalve diversity dynamics were explored. Aberhan and Fürsich (1997) proposed that the preferential disappearance of endemic bivalves at the early Toarcian extinction event in South America could be partly explained by immigration of cosmopolitan species via the corridor and subsequent competitive replacement. In addition, it was suggested that the rise in NW European bivalve diversity during Toarcian–Aalenian times was a direct consequence of taxa immigration through the corridor from South America, presumably filling the vacated ecospace after the early Toarcian extinction event (Hallam 1983; Hallam and Wignall 1997). Based on a detailed survey of time ranges at the species level for bivalves from each side of the corridor Aberhan (2002) concluded that neither of these hypothesis can be sustained, and suggested that the biotic recovery in both regions was largely controlled by increasing within-region origination rates rather than by immigration.

6.3   Oceanic Currents

The distribution of some Triassic and Jurassic bivalve species was also used as evidence to propose or favor certain ocean circulation models. Muller and Ferguson (1939) pointed out that Norian faunas of western North America were more similar to those from Mediterranean and especially Alpine regions than to those from the westernmost Tethys areas. To explain this pattern, Kristan-Tollmann and Tollmann (1981) envisaged a westerly directed "Tethyan current" for late Triassic times, and proposed such direction for the migration of the marine fauna. Some bivalve data are indeed in agreement with this hypothesis. For instance, during Triassic times *Palaeolopha* was common in eastern Panthalassa and central Europe, but was absent from Western Europe. The distribution of the peculiar late Triassic genus *Wallowaconcha*, abundant in eastern Panthalassa and also present in central and eastern Tethys (Arabia and south Asia), suggests westwards migration across Panthalassa (Yancey et al. 2005). This Triassic water-movement pattern also implies the absence of a direct connection between western Tethys and eastern Pacific at that time.

Bivalves were again used as arguments for the distribution of oceanic surface currents during the Jurassic, sometimes in a contradictory manner. For instance, while Nauss and Smith (1988) advocated a late Pliensbachian migration through the Hispanic Corridor for the main elements of the *Lithiotis* fauna, Krobicki and Golonka (2009) proposed an eastwards migration from western Tethys through Panthalassa up to the western margin of North and South America. This last proposal would mean the existence of a strong eastwards surface equatorial current, while the prevailing ideas favor an opposite-directed current for that time (see Sect. 1.6).

Data analyzed here strongly suggest that the Hispanic Corridor was the most probable migration route for benthonic bivalves. Nevertheless, when "neritic" bivalve species are analyzed separately from the benthonic ones, and although data are still few, their distribution during the Early Jurassic can be alternatively explained using a pantropic model, implying planktotrophic larvae and fast ocean currents. In this case, data agree better with a westwards direction of prevailing currents (in agreement with the proposal of Kristan-Tollmann and Tollmann 1981 for Late Triassic times), since some taxa which are abundant along the eastern Pacific, such as *Posidonotis* and *Otapiria*, do have isolated late records in eastern (but not western) Tethys (Fig. 6.4).

6.4 Evolution of Global Biogeographic Boundaries

As previously mentioned, transitional zones occurred between the Tethyan and
South Pacific first-order units. To study the evolution of these boundary zones
requires a very detailed set of data, which is not available for the key areas in the
Southern Hemisphere, except for a few, such as the already described ones along
the western South American margin (Sect. 4.3). Nevertheless, some broad features
will be pointed out, since it is interesting to compare them with the evolution of
Boreal/Tethyan boundary zones, which are somewhat better known. The evolution
of the boundary between Tethyan and Boreal Realms has been the subject of many
contributions, most of them based on the distribution of ammonite taxa (e.g. Imlay
1965; Hallam 1969, 1981, 1994; Fürsich and Sykes 1977; Hayami 1990;
Dommergues and Meister 1991; Hillebrandt et al. 1992), and on the basis of
bivalve distribution in Europe by Liu (1995).

For the northwest Pacific, the Tethyan–Boreal boundary based on bivalves was
between 36 and 34° during the Late Triassic in Japan (Tamura 1990) but was
attributed to accretion, and it was situated in the lower Amur region of Siberia
during the Early Jurassic (approximately 48° N) according to Hayami (1990). Sey
and Kalacheva (1985) suggested that the Boreal–Tethyan ammonite boundary was
about 60° N during the Sinemurian, a few degrees further north during the
Pliensbachian and reached about 75° N during Bajocian/Bathonian times.

For the northeast Pacific Taylor et al. (1984) record a northwards shift (about
8–10°) of the Boreal–Tehyan boundary and a 30° northwards shift in the reach of
east Pacific faunas between Sinemurian and Pliensbachian times. For the
Pliensbachian the boundary is marked by the extension of the *Lithiotis* reef faunas.
Dommergues and Meister (1991) also reported a northwards shift in the boundary
zone between Euroboreal and Tethyan ammonite faunas from Sinemurian to
Domerian times.

All data indicate that a northwards shift of the Boreal/Tethyan boundary from
Pliensbachian to Bathonian times took place, and that the Bathonian can be recognized as the time of the greatest Tethyan spread during the Jurassic. The
boundary then moved southwards during the Callovian (Liu 1995, p. 49), when the
distinctiveness of the Boreal bivalve fauna strongly increased and was maximum
at the Oxfordian (Liu 1995, p. 50). The Callovian and early Oxfordian marked the
widest spread of the Boreal Realm according to Hallam (1971).

From the possible transitional zones located in paleotemperate regions of the
Southern Hemisphere, the relationship between Maorian-type and other faunas in
the western Pacific was studied by Hayami (1984). In the eastern Pacific, the South
Andean boundary zone has been recognized for Early Jurassic times (Sect. 4.2).
Based on species distribution along the eastern Paleo-Pacific margin, a southwards
shift of this boundary of about 8–10° latitude has been proposed for the Hettangian–Toarcian interval. It is interesting to compare these observations with the
evolution of the boundary areas between Tethyan and Boreal Realms in the
Jurassic. To make this comparison, the boundary areas proposed on the basis of

Fig. 6.5 Position of the boundary zone between Tethyan and high-latitude first-order paleobiogeographic units based on bivalve distribution for the Pliensbachian and Toarcian. Sources discussed in text. Base map as in Fig. 1.2. (From Damborenea 2002b)

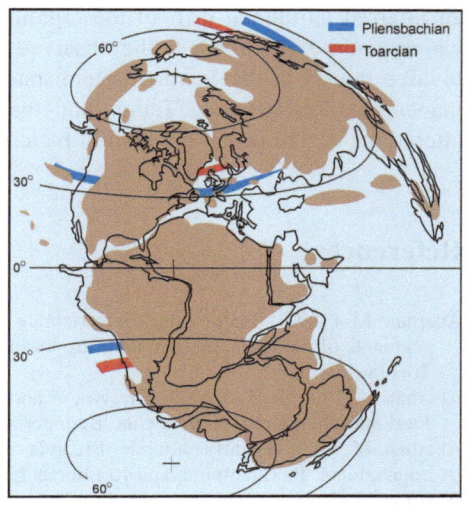

bivalve distribution (Hayami 1984, 1990; Smith and Tipper 1986; Smith 1989; Liu 1995; Damborenea 1996; Liu et al. 1998; Aberhan 1999), which are not always coincident with the limits proposed on the basis of ammonites, have been plotted on Fig. 6.5 for the Pliensbachian and Toarcian. The results show a high congruence between the behavior of the boundary in both hemispheres during the Early Jurassic: the southwards shift in the Southern Hemisphere is matched by an equivalent northward shift in the Northern Hemisphere. There are not yet detailed enough data from the Southern Hemisphere to follow this comparison for later Jurassic times. It is interesting to point out, though, that the greatest expansion of low-latitude bivalve faunas in the Southern Hemisphere occurred during Middle Jurassic times (see also Grant-Mackie et al. 2000), and that the northward shift of the Maorian/Tethyan boundary during Late Jurassic times in western Pacific regions, albeit somewhat complicated by the tectonic history of the area (Hayami 1984, 1987), roughly corresponds with the widest expansion of the Boreal Realm in the Northern Hemisphere. These coincidences suggest that the main faunal shifts are related in their origin.

One of the plausible explanations which can account for both the northern and southern symmetric migration of the boundaries of the Tethyan Realm toward the poles during the Pliensbachian and Toarcian is a worldwide warming of the climate at that time. Based on the study of the distribution of Holocene mollusks in the Sea of Japan, Lutaenko (1993) concluded that an increase of 0.1–0.2° C in surface water temperature can cause a shift of about 100 km of the boundary between warm water mollusks and the Pacific boreal region.

Of course, climatic changes are often invoked to explain such shifts in paleobiogeographic boundaries. However, it is far more difficult to discuss what the cause of such supposed climatic changes was. The main paleogeographic events which should have influenced the climate of this area during the Jurassic were: (a) an inferred latitudinal drift of the region concerned (Iglesia-Llanos et al. 2006), which could explain *per se* the observed displacement of climatically controlled bivalve faunas in the Southern Hemisphere, and (b) the establishment of a permanent seaway between Tethys and the eastern Pacific, producing large-scale alterations on the ocean circulation patterns and climate.

References

Aberhan M (1993) Benthic macroinvertebrate associations on a carbonate-clastic ramp in segments of the Early Jurassic back-arc basin of northern Chile (26–29° S). Rev Geol Chile 20:105–136

Aberhan M (1994) Early Jurassic Bivalvia of northern Chile. Part I. Subclasses Palaeotaxodonta, Pteriomorphia, and Isofilibranchia. Beringeria 13:1–115

Aberhan M (1998a) Early Jurassic Bivalvia of western Canada. Part I. Subclasses Palaeotaxodonta, Pterionorphia, and Isofilibranchia. Beringeria 21:57–150

Aberhan M (1998b) Paleobiogeographic Patterns of Pectinoid Bivalves and the Early Jurassic Tectonic evolution of Western Canadian Terranes. Palaios 13:129–148

Aberhan M (1999) Terrane history of the Canadian Cordillera: estimating amounts of latitudinal displacement and rotation of Wrangellia and Stikinia. Geol Mag 136:481–492

Aberhan M (2001) Bivalve palaeobiogeography and the Hispanic corridor: time of opening and effectiveness of a proto-Atlantic seaway. Palaeogeogr Palaeoclimatol Palaeoecol 165:375–394

Aberhan M (2002) Opening of the Hispanic corridor and Early Jurassic bivalve biodiversity. In: Crame JA, Owen AW (eds) Paleobiogeography and biodiversity change: the Ordovician and Mesozoic-Cenozoic radiation. Geol Soc London Spec Publ 194:127–139

Aberhan M, Fürsich FT (1997) Diversity analysis of lower Jurassic bivalves of the Andean Basin and the Pliensbachian/Toarcian mass extinction. Lethaia 29:181–195

Aberhan M, Hillebrandt A (1999) The bivalve *Opisoma* in the lower Jurassic of northern Chile. Profil 16:149–164

Arp G, Seppelt S (2012) The bipolar bivalve *Oxytoma* (*Palmoxytoma*) *cygnipes* (Young & Bird, 1822) in the Upper Pliensbachian of Germany. Paläontol Z 86:43–57

Broglio Loriga C, Neri C (1976) Aspetti paleobiologici e paleogeografici della facies a "*Lithiotis*" (Giurese inf.). Riv Ital Paleontol Stratigr 82:651–705

Campbell HJ (1994) The Triassic Bivalves *Daonella* and *Halobia* in New Zealand, New Caledonia, and Svalbard. Inst Geol Nucl Sci Mon 4:1–166

Cecca F (2009) La dimension biogéographique de l'évolution de la Vie. CR Palevol 8:119–132

Crame JA (1981) The occurrence of *Anopaea* (Bivalvia: Inoceramidae) in the Antarctic Peninsula. J Mollusc Stud 47:206–219

Crame JA (1986) Late Mesozoic bipolar bivalve faunas. Geol Mag 123:611–618

Crame JA (1987) Late Mesozoic bivalve biogeography of Antarctica. Proc Sixth Gondwana Symp (Columbus, Ohio):93–102

Crame JA (1992) Evolutionary history of the polar regions. Hist Biol 6:37–60

Crame JA (1993) Bipolar molluscs and their evolutionary implications. J Biogeogr 20:145–161

Crame JA (1996a) Evolution of high-latitude molluscan faunas. In: Taylor JD (ed) Origin and evolutionary radiation of the Mollusca. Oxford University Press, Oxford

Crame JA (1996b) Antarctica and the evolution of taxonomic diversity gradients in the marine realm. Terra Antarct 3:121–134

Damborenea SE (1990) Middle Jurassic inoceramids from Argentina. J Paleontol 64:736–759

Damborenea SE (1993) Early Jurassic South American pectinaceans and circum-Pacific palaeobiogeography. Palaeogeogr Palaeoclimatol Palaeoecol 100:109–123

Damborenea SE (1996) Palaeobiogeography of Early Jurassic bivalves along the southeastern Pacific margin. 13° Congr Geol Argent y 3° Congr Explor Hidrocarb (Buenos Aires). Actas 5:151–167

Damborenea SE (2000) Hispanic corridor: its evolution and the biogeography of bivalve molluscs. In: Hall RL, Smith PL (eds) Advances in Jurassic research 2000. GeoResearch Forum, vol 6. pp 369–380

Damborenea SE (2002a) Early Jurassic bivalves from Argentina. Part 3: Superfamilies Monotoidea, Pectinoidea, Plicatuloidea and Dimyoidea. Palaeontographica A 265:1–119

Damborenea SE (2002b) Jurassic evolution of Southern Hemisphere marine palaeobiogeographic units based on benthonic bivalves. Geobios 35 MS 24:51–71

Damborenea SE (2004) Early Jurassic Kalentera (Bivalvia) from Argentina and its palaeobiogeograhical significance. Ameghiniana 41:185–198

Damborenea SE, González-León CM (1997) Late Triassic and Early Jurassic bivalves from Sonora, Mexico. Rev Mex Cienc Geol 14:178–201

Damborenea SE, Manceñido MO (1979) On the palaeogeographical distribution of the pectinid genus Weyla (Bivalvia, Lower Jurassic). Palaeogeogr Palaeoclimatol Palaeoecol 27:85–102

Damborenea SE, Manceñido MO (1988) Weyla: semblanza de un bivalvo Jurásico andino. Actas 5° Congr Geol Chileno 2: C13–C25

Damborenea SE, Manceñido MO (1992) A comparison of Jurassic marine benthonic faunas from South America and New Zealand. J Roy Soc N Z 22:131–152

Damborenea SE, Manceñido MO (2012) Late Triassic bivalves and brachiopods from southern Mendoza, Argentina. Rev Paléobiol VS 11:317–344

Darwin C (1859) The origin of species by means of natural selection, or the preservation of favoured races in the struggle for life. John Murray, London

Dhondt AV (1992) Cretaceous inoceramid biogeography: a review. Palaeogeogr Palaeoclimatol Palaeoecol 92:217–232

Dommergues JL, Meister C (1991) Area of mixed marine faunas between two major paleogeographical realms, exemplified by the Early Jurassic (Late Sinemurian and Pliensbachian) ammonites in the Alps. Palaeogeogr Palaeoclimatol Palaeoecol 86:265–282

Donoghue MJ (2011) Bipolar biogeography. Proc Nat Acad Sci 108:6341–6342

Frebold H (1957) The Jurassic Fernie group in the Canadian Rocky Mountains and foothills. Memoir Geol Soc Surv Canada 287:1–197

Fürsich FT, Sykes RM (1977) Palaeobiogeography of the European Boreal realm during Oxfordian (Upper Jurassic) times: a quantitative approach. N Jb Geol Paläontol Abh 172:271–329

Grant-Mackie JA, Aita Y, Balme BE, Campbell HJ, Challinor AB, MacFarlan DAB, Molnar RE, Stevens GR, Thulborn RA (2000) Jurassic palaeobiogeography of Australasia. In: Wright AJ, Young GC, Talent JA, Laurie JR (eds) Palaeobiogeography of Australasian faunas and floras. Memoir of the Association of Australasian Palaeontologists, vol 23. pp 311–353

Hallam A (1969) Faunal realms and facies in the Jurassic. Palaeontology 12:1–18

Hallam A (1971) Provinciality in Jurassic faunas in relation to facies and palaeogeography. In: Middlemiss FA, Rawson PF, Newall G (eds) Faunal provinces in space and time. Geol J Spec Issue 4:129–152

Hallam A (1977) Jurassic bivalve biogeography. Paleobiology 3:58–73

Hallam A (1981) Relative importance of plate movements, Eustasy, and climate in controlling major biogeographical changes since the Early Mesozoic. In: Nelson G, Rosen DE (eds) Vicariance biogeography: a critique. Columbia University Press, New York

Hallam A (1983) Early and mid-Jurassic molluscan biogeography and the establishement of the central Atlantic seaway. Palaeogeogr Palaeoclimatol Palaeoecol 43:181–193

Hallam A (1994) Jurassic climates as inferred from the sedimentary and fossil record. In: Allen J, Hoskins B, Sellwood B, Spicer R, Valdes P (eds) Palaeoclimates and their modelling with special reference to the Mesozoic Era. Chapman and Hall, London

Hallam A, Wignall PB (1997) Mass extinctions and their aftermath. Oxford University Press, Oxford

Hayami I (1975) A systematic survey of the Mesozoic Bivalvia from Japan. Bull Univ Mus Univ Tokyo 10:1–249

Hayami I (1984) Jurassic Marine Bivalve Faunas and biogeography in Southeast Asia. Geol Palaeontol SE Asia 25:229–237

Hayami I (1987) Geohistorical background of Wallace's line and Jurassic marine biogeography. In: Taira A, Tashiro M (eds) Historical biogeography and plate tectonic evolution of Japan and Eastern Asia. , Tokyo

Hayami I (1990) Geographic distribution of Jurassic Faunas in Eastern Asia. In: Ichikawa K, Mizutani S, Hara I, Hada S, Yao A (eds) Pre-cretaceous terranes of Japan. Publication of IGCP Project 224, Osaka

Hillebrandt A (1981) Kontinentalverschiebung und die paläozoogeographischen Beziehungen des südamerikanischen Lias. Geol Runds 70:570–582

Hillebrandt A, Westermann GEG, Callomon JH, Detterman RL (1992) Ammonites of the Circum-Pacific region. In: Westermann GEG (ed) The Jurassic of the Circum-Pacific. Cambridge University Press, New York

Iglesia-Llanos MP, Riccardi AC, Singer SE (2006) Palaeomagnetic study of lower Jurassic marine strata from the Neuquén Basin, Argentina: a new Jurassic apparent polar wander path for South America. Earth Planet Sci Let 252:379–397

Imlay RW (1965) Jurassic marine faunal differentiation in North America. J Paleontol 39:1023–1038

Jaworski E (1929) Eine Lias-Fauna aus Nordwest-Mexiko. Abh Schweizer Palaeontol Ges 48:1–12

Kauffman EG (1973) Cretaceous Bivalvia. In: Hallam A (ed) Atlas of Palaeobiogeography. Elsevier, Amsterdam

Kelly SRA (1984) Bivalvia of the Spilsby Sandstone and Sandringham Sands (Late Jurassic–Early Cretaceous) of eastern England. Part I. Palaeontogr Soc Mon 137(566):1–100

Kristan-Tollmann E, Tollmann A (1981) Die Stellung der Tethys in der Trias und die Herkunft ihrer Fauna. Mitt österreich Gesells 74–75:129–135

Krobicki M, Golonka J (2009) Palaeobiogeography of Early Jurassic Lithiotis-type bivalve buildups as recovery effect after Triassic/Jurassic mass extinction and their connection with Asian palaeogeography. Acta Geoscient Sin 30(supl. 1):30–33

Liu C (1995) Jurassic bivalve palaeobiogeography of the Proto-Atlantic and the application of multivariate analysis methods in palaeobiogeography. Beringeria 16:3–123

Liu C, Heinze M, Fürsich FT (1998) Bivalve provinces in the Proto-Atlantic and along the southern margin of the Tethys in the Jurassic. Palaeogeogr Palaeoclimatol Palaeoecol 137:127–151

Lutaenko KA (1993) Climatic optimum during the Holocene and the distribution of warm-water mollusks in the Sea of Japan. Palaeogeogr Palaeoclimatol Palaeoecol 102:273–281

McRoberts CA (1997) Late Triassic (Norian–Rhaetian) bivalves from the Antimonio Formation, northwestern Sonora, Mexico. Rev Mex Cienc Geol 14(2):167–177

Milova LV (1988) Ranneyurskie dvustvorchatye Mollyuski Severo-Vostoka SSSR. Akad Nauk SSSR, Vladivostok

Muller S, Ferguson H (1939) Mesozoic Stratigraphy of the Hawthorne and Tonopah Quadrangles Nevada. Bull Geol Soc Am 50(10):1573–1627

Nauss AL, Smith PL (1988) Lithiotis (Bivalvia) bioherms in the Lower Jurassic of East-central Oregon, U.S.A. Palaeogeogr Palaeoclimatol Palaeoecol 65:253–268

Newton CR (1988) Significance of "Tethyan" fossils in the American Cordillera. Science 242:385–391

Riccardi AC (1991) Jurassic and Cretaceous marine connections between the Southeast Pacific and Tethys. Palaeogeogr Palaeoclimatol Palaeoecol 87:155–189

Ros S, Márquez-Aliaga A, Damborenea SE (2012) Comprehensive database on Induan (Lower Triassic) to Sinemurian (Lower Jurassic) marine bivalve genera and their paleobiogeographic record. Paleontol Contrib Univ Kansas (in press)

Sandoval J, Westermann GEG (1986) The Bajocian (Jurassic) ammonite fauna of Oaxaca, Mexico. J Paleontol 60:1220–1271

Sandy MR, Stanley GD (1993) Late Triassic brachiopods from the Luning formation, Nevada, and their paleogeographocal significance. Palaeontology 36:439–480

Scholz A, Aberhan M, González-León CM (2008) Early Jurassic bivalves of the Antimonio terrane (Sonora, NW Mexico): taxonomy, biogeography, and paleogeographic implications. Geol Soc Am Spec Pap 442:269–312

Sey II, Kalacheva ED (1985) Invasions of Tethyan ammonites into the Late Jurassic Boreal basins of East U.S.S.R. In: Westermann GEG (ed) Jurassic biogeography and stratigraphy of East USSR. IGCP Project 171: Circum-Pacific Jurassic, Special Paper, vol 10. pp 14–17

Sey II, Polubotko IV (1992) Atlas, Pl. In: Westermann GEG (ed) The Jurassic of the Circum-Pacific. Cambridge University Press, Cambridge, pp 120–127

Sha J (1996) Antitropicality of the Mesozoic Bivalves. In: Pang ZH et al (eds) Advances in solid earth sciences. Science Press, Peking

Sha J (2002) Hispanic Corridor formed as early as Hettangian: on the basis of bivalve fossils. Chinese Sci Bull 47:414–417

Sha J (2003) Plankton and pseudoplankton of the marine Mesozoic bivalves. Acta Paleontol Sin 42:408–416

Shi GR, Grunt TA (2000) Permian Gondwana-Boreal antitropicality with special reference to brachiopod faunas. Palaeogeogr Palaeoclimatol Palaeoecol 155:239–263

Smith PL (1989) Paleobiogeography and plate tectonics. Geosci Canada 15:261–279

Smith PL, Tipper HW (1986) Plate tectonics and paleobiogeography: Early Jurassic (Pliensbachian) endemism and diversity. Palaios 1:399–412

Smith PL, Westermann GEG, Stanley GD Jr, Yancey TE (1990) Paleobiogeography of the Ancient Pacific (response by Newton, C.R.). Science 249:680–683

Stanley GD Jr, González-León CM (1997) New late Triassic scleractinian corals from the Antimonio Formation, northwestern Sonora, Mexico. Rev Mex Cienc Geol 14:202–207

Stanley GD Jr, González-León C, Sandy MR, Senowbari-Daryan B, Doyle P, Tamura M, Erwin DH (1994) Upper Triassic invertebrates from the Antimonio Formation, Sonora, Mexico. Paleontol Soc Mem 36 (Suppl J Paleontol 68):1–33

Tamura M (1990) The distribution of Japanese Triassic bivalve funas with special reference to parallel distribution of inner Arcto-Pacific fauna and outer Tethyan fauna in Upper Triassic. In: Ichikawa K, Mizutani S, Hara I, Hara S, Yao A (eds) Pre-Cretaceous terranes of Japan. Publ. IGCP Project, vol 224. pp 347—359

Taylor DG, Callomon JH, Hall R, Smith PL, Tipper HW, Westermann GEG (1984) Jurassic ammonite biogeography of western North America: the tectonic implications. In: Westermann GEG (ed) Jurassic-Cretaceous Biochronology and Paleogeography of North America. Geol Assoc Canada Spec Pap 27:121–141

Tozer ET (1982) Marine Triassic faunas of North America: their significance for assessing plate and terrane movements. Geol Runds 71:1077–1104

Yancey TE, Stanley GD Jr, Piller WE, Woods MA (2005) Biogeography of the Late Triassic wallowaconchid megalodontoid bivalves. Lethaia 38:351–365

Zinsmeister WJ (1982) Late Cretaceous-Early Tertiary molluscan biogeography of the southern Circum-Pacific. J Paleontol 56:84–102